W9-AXP-943

# WHERE THERE'S SMOKE...

## How to protect yourself, your family your property, your environment from fire!

by Robin F. Pendergrast

Dorison House Publishers     Boston

## The Author

Robin F. Pendergrast is the captain of the Northfield Rescue Squad, Northfield, Illinois. In addition he is a volunteer firefighter with the Northfield Fire Department, and before that with the Winnetka, Illinois, Fire Department. A long-time advocate of smoke detectors, he coordinated a comprehensive smoke detector program for the village of Northfield.

A recent project of his made the NBC "good" news. It was a presentation of a fire engine to the leper colony, Kalaupapa, in Hawaii. His enthusiasm for fire safety knows no bounds.

All rights reserved including the right of reproduction in whole or in part in any form.

Copyright ©1982 by Dorison House Publishers, Inc.

Published by Dorison House Publishers, Inc.
824 Park Square Building, Boston, Massachusetts   02116

ISBN: 916752-53-4
Library of Congress Catalog Card Number: 81-69076

Manufactured in the United States of America

Book design by Myra Lee Conway/Design;
Illustrations by Diane Jaquith.

# CONTENTS

# THERE'S FIRE!

## "A Gift of The Gods"

## Mythology of Fire

Fire has a long and noble history. Although we often think of fire today as a scourge that takes lives and destroys property, we must remember that it is also humanity's greatest tool. We have advanced so far down the path of civilization and technology that we sometimes forget the harnessing of fire was one of mankind's greatest accomplishments.

Fire provided primitive man with his first heating source. Fire opened up the night, allowing people to see their way in the dark. Fire purified metal and so led to superior tools. Primitives who hunted herds of game would deliberately start brush fires to drive prey into traps.

Ancient civilizations recognized the importance of fire and draped ceremony around it.Gods of old Athens and Rome like Zeus and Vulcan were worshipped for their connection with fire (the one the thunderbolt thrower, the other the smith). The ancient Greek sage Heraclitus saw fire as the primal element in nature. Peasants looked upon their hearth fire as having a divine presence.

Fire has always been a symbol of both continuity and of new beginnings. In today's world, we speak of handing the torch of past generations to the new one. The hearths of new colonies planted by the ancient Greeks were lit with fire brought by ship from the mother city. Every fourth year we see the ignition of an Olympic torch symbolizing athletic events that took place 2,500 years ago.

The idea of making use of fire may have come to man by observation of brush fires or of lightning striking trees. Early peoples learned to create fire themselves from rubbing wood together or using flint. Despite the obvious connection of the material world to fire, man surrounded the original discovery of fire in myth. In the South Seas, for example, legend says that fire comes from a goddess. The goddess set everything on fire, one time long ago, by producing flames through her fingers and toes. Rainfall extinguished most of the conflagration, the story goes, but some fire lingered inside trees. The colorful myth ends with the conclusion that people have been able to obtain fire from vegetation ever since (by rubbing sticks together).

## History of Fire Protection

Fire protection has an interesting history too. As Civilization became more complex and people began to live together in cities, fire grew as a danger. Urban living meant that homes were packed closely together or on top of one another. A fire in one home could become a threat to all the rest.

Communities began to pass laws and develop techniques to protect themselves from the threat of fire.

After he had invaded England in 1066, William the Conqueror became concerned at the disastrous fires resulting from the open hearths kept in the homes of his subjects. He established a law that stipulated a fire cover be placed over hearths each evening. The cover stopped

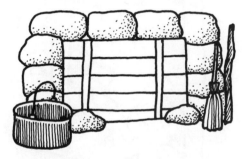

sparks from setting ablaze the interior of a home. Two centuries later in England, fires were commonly put out by pulling the corner posts out from beneath buildings. Town aldermen were required to have iron chains in storage to do this.

One of the first fire departments was formed by the Roman Emperor Augustus about 2,000 years ago. The department was composed of squads of men commanded by a fire chief who rode to blazes in his chariot. The firemen were organized into groups mimicking the military organizations of the day.

According to law, a sort of fire marshal investigated the cause of each fire.

Fire carts equipped with flexible hoses and hand pumps made their appearance in Holland in the 1670s. In colonial America at about the same time, fire was a dreaded threat to the advancing civilization. It wiped out towns and leveled whole sections of towns and cities. Fire laws were passed against citizens who failed to keep their chimneys clean, and volunteer fire companies were organized to fight back.

The shift to paid departments came gradually. Boston was the first to establish one after a great conflagration in 1679, but for the next 200 years separate volunteer companies survived in most cities.

**Improvements In Fire Protection**
About a century ago the battle against fire became more of a science. Significant advances were made toward better fire protection by the analysis of what caused fires. Insurance companies with a financial stake in preventing damage to insured property led the way in formulating building code restrictions.

The most famous fire in U.S. history, the Chicago Fire of 1871, served to spur reform efforts. (The Chicago Fire was abetted less by Mrs. O'Leary's cow than by a fire watcher who mistakenly judged the fire's location and sent fire companies to the wrong place.) The loss of about 120 lives to the inferno led to the establishment of Fire Prevention Week.

Throughout this period technology was developing better equipment to fight fire. Around 1850 steam-powered fire engines replaced hand pumps. Sixty years later the gasoline-powered fire engine made its appearance. "Fly ladders" 55 feet in length were developed which allowed firemen to rescue people trapped many floors off the ground.

Near the turn of the century the still new source of power called electricity came to be recognized as a potential fire hazard. An engineer named William Merrill convinced insurance companies to set up a testing laboratory, Underwriters' Laboratories, in 1894. Five years before, the National Fire Protection Association was formed to set standards for fire protection equipment like fire doors and fire extinguishers.

**Modern Fire Control**

Decades of research and clever application of technology has led to modern methods of controlling fire. For example, fire-resistant materials have been developed which can stand up to fire for relatively long periods of time. Wood can be rendered less flammable by treating it with such substances as phosphates. Wooden buildings, traditionally firetraps, are made safer by the addition of fire-resistant sidings. Walls constructed of steel and asbestos are very fire-resistant, and fire walls between and within buildings along with fire floors, contain fire within a compartment until it can be put out.

Today you can treat clothing and draperies with fire-retardant chemical spray. Even Christmas trees can be treated with fire-retardants.

In the sixties, buildings began filling up with plastics which have become a new and dangerous fuel. Research into the complex physical and chemical dynamics of fire is being conducted using the latest computers. New, scientifically-based fire codes are likely to result

But as useful as fire science may be in assuring our future safety, probably the most important breakthrough in home life saving devices is the early-warning detector. 75 percent of all fires, deaths and injuries occur in the family homes and apartments. Most Americans who died in home fires were overcome by smoke, toxic gases, or lack of oxygen.

In countless instances these lives would have been saved if the victims had been awakened to the presence of a fire in its early stages. Experts say, and the figures show, that fire losses could be cut almost in half in a single year if there were no arsonists or cigarette smokers, **and if every home had smoke alarms.**

Other significant developments in fire protection have come in the field of extinguishment. Sprinkler systems and modern fire departments provide ways of fighting fire undreamed of in the days of the horse-drawn pumper and the bucket brigade.

Automatic sprinkler systems, statistics prove, offer remarkably effective fire protection. Sprinklers

bring under control over 95 percent of the fires starting in the area of the system. When sprinklers fail to stop fire the fault often lies with a deficient supply of water.

Sprinklers work so well because they literally flood a fire before it can get started. No other fire protection device responds so fast. A system puts out as much as 1,500 gallons of water each minute! Good systems have at least one backup water source in the event that water cannot be obtained from the main source.

Heat sensors in the sprinklers serve to set off the system. Sensors can be built so that only very high temperatures cause a flow of water: a factory with equipment that gives off a great deal of heat might contain these kinds of sensors.

Business managers sometimes refuse to install sprinklers, objecting to the cost and the chance that sprinklers can cause water

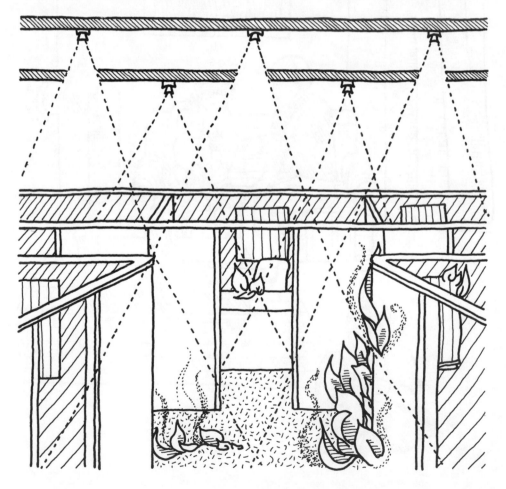

damage. Needless to say, the cost of a fire is liable to be greater than any harm done by water. In addition, there are sprinkler systems in which one or two prelimininary actions take place before water is actually released. For example, an alarm may first be set off to give workers a chance to put out a fire with fire extinguishers. An interval of time passes before the room or storage area is sprayed.

An unchanging conditon of battling fire is that fire personnel must put their lives on the line in performing their jobs. There are few things more dangerous than entering a burning building, an act fire fighters are called upon to do constantly. However, the modern fire fighter holds a huge technological edge over the people who made up the fire brigades at the time of the Chicago Fire.

Nowadays calls are dispatched to the fire station by computer. Fire fighters race to the scene in huge fast trucks instead of by chariot. Powerful pumpers allow a tremendous quantity of water to be dumped into a burning building. Rescuers going close to flames have the protection of heat-resistant clothing, and can communicate with each other in the worst noise and confusion through the radio sets built into their gas masks.

Those who do battle against forest fires wield even more spectacular equipment. Fire fighters parachute into burning woods from airplanes which fly on to make water drops.

The same aircraft can serve as long-range scouts, spotting the outbreak of trouble dozens of miles from human habitation. On the ground bulldozers and high-speed saws are used to cut treeless paths in the line of advancing firestorms.

### Fire for Warmth, Light and Comfort

In a now distant past, mythology pictured fire as a boon to mankind. At the moment in the long ago when mankind first put fire to work, it was the start of the march to civilization.

Uncontrolled, it was often a curse. Today, we can control fire and once again consider it a "gift of the gods."

### About This Book

It's important for every person to be well-informed about fire safety anywhere, anytime. With preparedness comes confidence, a feeling of security, and a calm attitude — all essentials in time of crisis.

The purpose of this book is to provide you with that security. You should be able to apply some of the information learned and technology developed from centuries of fire fighting to your own home and family.

There are three basic steps to take in order to practice good family safety:

1. Eliminate any hazardous conditions that can lead to ignition.

2. Install and maintain smoke detectors.

3. Be prepared to take effective

action to protect life and property should a fire break out.

## Don't Play With Fire

Everyone has heard the statement, "That's like playing with fire," countless times. The warning, along with many others like it is meant to get across the point that **actually** playing with fire can be dangerous. In fact, most fires are caused by carelessness — tolerating or directly contributing to hazardous conditions. They are commonly the result of such things as frayed electrical wires left unchecked, oil tank leaks gone unnoticed, and, yes, matches left carelessly within the playful reach of young children.

In the chapters that follow, we will point out the hazards found in the typical home, and what you can do about them, what you should check for when building, buying or renting a home, and how you can make the world safer for yourself and others wherever you are. Chapter VI gives you insurance indemnity facts, so you can be sure you have enough insurance coverage.

## Early Warning Saves Lives

To fight or escape from a fire, you need to know one exists. Yet your eyes, ears, and nose cannot be everywhere all the time, and especially not at night — the time when most home fires start. You need electronic watchdogs to set off a loud alarm when the first hint of fire is detected — which can be hours before anyone would be aware of the trouble without them.

This book will discuss in Chapter III the most effective use of smoke detectors and other fire detection and fighting equipment.

## Know-how Is The Best Defense

Recognizing the frequent occurence of fire — hundreds of thousands of separate incidents each year in the United States alone — it's apparent that fires can happen to even the most careful, fire-conscious people. Knowledge of proper "fire reaction" — how to fight a fire and escape from it — can make the difference between life and death. So, too, can a familiarity with emergency first-aid procedures.

*Where There's Smoke...* is a consciousness raising book. It will tell you how you can greatly reduce the chances of fire destroying your life.

# WHERE'S THE FIRE?

# Chapter I

Eleven Major
Causes of Fire
and What to Do
About Them

1 Careless Smoking Habits

2 Young Children

3 Ignited Rubbish

4 Combustible Solids

5 Inflammable Liquids

6 Faulty Electric Wiring

7 Malfunctioning Electric Appliances

8 Kitchen Cooking Fires

9 Improper Use of Heating Units

10 Uncontrolled Open Flames

11 Thawing Frozen Water Pipes

## IT'S UP TO YOU

You have it in your power to control fire.

Most fires can be prevented. It is clear from extensive research into the causes of fire that innumerable deaths and injuries and billions of dollars of property loss should never have taken place.

Seventy percent of the fires that occur in homes and other buildings can be attributed to carelessness. In other words, just by taking a few precautions, the large majority of fires can be eliminated.

Most other fires are the result of faulty equipment or appliances. In many cases, the defects should have been recognized and repaired before being allowed to spark a conflagration.

## ELEVEN MAJOR CAUSES OF FIRE

### 1. Careless Smoking Habits

The number one cause of home fires, lack of care with cigarettes, cigars and pipes, should be the easiest hazard to eliminate. As a danger to your health, smoking is to be strongly discouraged, but since some people do smoke, there must be very strict rules about the proper use and disposal of smoking materials.

Smoking should be limited to spacious, uncluttered rooms. Smoking in closed-in areas like attics and garages should be **verboten**

Never light a match or puff on a cigarette in a place containing

inflammable liquids like paint thinners, or other combustibles like paper, leaves, or rags. A lit match could set off the liquids as well as the combustible gases improperly stored liquids can produce.

Smoking in bed or in an easy chair carries more than the risk of the smoker falling asleep before putting out a pipe or cigar. A stray ember could fall unnoticed, even from an awake, alert smoker, and lodge itself deep within the fabric to burn slowly for hours before reaching flashpoint. Avoid this by never smoking around combustible material of any kind.

Alcohol and smoking make a particularly dangerous combination. Drinking tends to make people more careless about their smoking habits. Also, it is harder for someone who has been drinking to make an escape from a fire.

The combination can be especially dangerous at parties where drinking is sometimes excessive and where there is likely to be smoking. So be on the alert when you have friendly get-togethers. Make sure there are plenty of deep ashtrays placed about the smoking area. Empty the ashtrays regularly and, in doing so, make sure that all cinders and lit butts have been put out. To be extra safe, dispose of all smoking material in a separate, metal trash container.

When all the guests have left, clean up the rooms where people were socializing. Don't put clean-up off until the next morning.

Make a survey of the furniture and floors for stray ashes or matches.

## 2. Young Children

Young children constitute a "fire hazard" for two reasons:

1) They tend to start fires because they do not know any better.
2) Once a fire is under way, they lack the knowledge or experience to react correctly. They often cannot handle themselves in a stressful situation.

### Children Need Supervision

It may seem "safe" to leave children alone while they're playing, sleeping or watching TV, but it is not. The natural curiosity of young children leads them to do all sorts of dangerous things like climbing up on furniture, taking inedible objects into their mouths, and playing with

matches. If a fire starts when they're sleeping, they're helpless.

A number of studies have been done to determine the number and types of burns that take place among certain age groups. The research findings of the U.S. Consumer Products Safety Commission found that young children who have been burned frequently fail to understand how burns come about — contesting the proverb, "Burnt child fire dredth."

Further, the studies showed open flames as a primary cause of burns, typically involving children playing with matches or around stoves without adult supervision. Boys are burned 2½ times more often than girls, and

boys at 7 and 8, according to the report, suffered a great many burns from inflammable liquids. An overriding factor in these accidents was the lack of the presence of an adult.

Protect your children from fire and themselves. — A dependable babysitter should be hired, if you are going out. The babysitter should:

- know the relevant emergency phone numbers for the area, as well as the number where you can be reached in an emergency.
- understand the procedure for exiting the house with the children in the event of a fire.
- check on the children often after they've gone to bed.

### Children and Fireworks

Fireworks are banned in many states. If they are permitted in your locality, and if you decide to let your children use them, make sure you are present. Inspect rockets and firecrackers personally before allowing your children to use any of them. Fireworks should only be lit with a flameless, slow-burning torch. Matches are dangerous because a flickering flame may light the lower end of a fuse, causing an explosion sooner than expected.

### Halloween Trick-or-Treating

Children should carry flashlights, not candles, when they go trick-or-treating. Costumes should be flame-retardant.

### Christmas Trees

Avoid using inflammable decorations on or around the tree. Do not play with toy trains, motors, or engines under or near the tree. Keep holidays happy.

**They've Got To Be Taught**
As soon as they are old enough to understand, children should be taught not to play with matches, stoves and extension cords.

The childhood years are the best time to instill habits of fire safety, especially since youngsters are particularly prone to fire accidents.

**In the Home**
The saddest thing in the world is to imagine a child who hears a smoke detector alarm go off and doesn't know what it is or what to do. Everyone in the household should know where equipment to detect and fight fires is in the home and how to operate it. (See pages 63-73.)

Children can and must learn fire awareness. See pages 89-98 for a detailed description of Operation EDITH (Exit Drills In The Home) which encourages families to devise — and rehearse — plans for getting the family out of the house in the event of a fire.

**In the Schools**
Fire safety eduction programs in schools have proven to be very effective. Sparky, the Fire Dog, a creation of the National Fire

Protection Association, has been teaching fire safety to children in schools all over the country. Some communities have exemplary fire safety education programs, notably, Santa Ana, California. This city of 165,000 people has an imaginative program in the classrooms, supplemented by demonstrations by the fire department, a parade at the end of Fire Prevention Week, a poster contest, and a carnival. Civic groups, and the fire department have cooperated with the schools in this program.

But Santa Ana is unusual, and the National Commission on Fire Prevention and Control has recommended that the Department of Health, Education, and Welfare include in accreditation standards fire safety education in the schools throughout the school year. Only schools presenting an effective fire safety education program would be eligible for any Federal financial assistance.

Local fire departments make significant contributions to public education through inspections of dwelling and commercial establishments, distribution of reading material on fire safety, and through cooperation with schools. Many fire departments encourage children to visit them, examine the fire fighting apparatus, and become involved in planning the family escape plan. (See page 90.)

### 3. Ignited Rubbish

The extra emphasis on this source of fire is intentional. Be particularly careful of matches or cigarettes outside the home as well as inside.

A carelessly discarded match, cigarette, cigar, or even pipe ashes is one of the major ways to start a fire in leaves, grass or paper. Frequently it happens in a garage or shed where inflammable liquids and solids are stored.

Small, perhaps even decorative, buckets of sand strategically placed, can help eliminate this problem.

### 4. Combustible Solids

An accumulation of oily rags, piles of newspapers, heaps of wood blocks and shavings in closets, basements, and hidden cubby-holes is a primary cause of fires, as they can be breeding grounds for fires or add a great deal of tinder to already existing fires. They can also block escape passages if there is a fire.

The remedy here is obvious: don't let trash pile up. Check the house from time to time to find out if unneeded items have been safely thrown away.

### 5. Inflammable Liquids

Thoughtless handling and storage of inflammable liquids is another major cause of fires and casualities. Inflammable liquids include cleaning fluids, paints, oil, insect sprays, gasoline, and benzene. These substances must be stored inside approved metal safety containers in well-ventilated areas. The containers should be leak-free and neatly arranged to prevent someone from accidently tripping over them. Pay particular attention to cleaners stored in cabinets under

kitchen sinks. If you have no forseeable need for a certain cleaner or spray, throw it away. (Make sure you wash thoroughly if any of these substances comes in contact with your skin.)

Spilled gasoline evaporates into the air quickly. It is the vapors produced by inflammable liquids, not the liquids themselves, that actually burn. The gases have to mix with oxygen in the air to ignite. This is why it is so important that lids or caps be securely positioned onto containers.

Gasoline, benzene, or naptha are not meant to be used as cleaning solvents. Their flashpoints are too low for safe use. Also, never carry gasoline, or any inflammable liquid, in the trunk of a car. They are too explosive to be carried around in a moving vehicle since vapors could leak, and in case of accident, it could be disastrous. To transport gasoline from the station, be sure cap is tight and have container only ¾ full to relieve expansion pressure.

Store lawn mowers and all gas-powered machines empty; refuel out of doors.

### 6. Faulty Electrical Wiring

Some people have the wrong-headed idea that they are powerless to prevent a fire set off by a spark from a defective wire. Such a malfunction can take place completely out of sight or in the middle of the night, goes this poor line of reasoning, and so nothing can be done. In actuality,

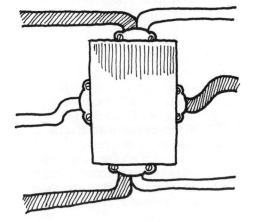

you can avoid trouble with a bit
of preventive maintenance.

Either check out the wiring
yourself or, if terms like "ohm" and
"cathode" are foreign to you, get
a qualified electrician to spend
an hour or two inspecting the
house's wiring. Regular
inspections are a lifesaving must,
especially so in older buildings
where wiring may have been in
place for fifty or sixty years.

A professional can tell
whether or not wires need to be
replaced. Frayed, old lines should
be replaced immediately.

### Circuit Breakers And Fuses: A Line of Defense

The normal operation of an
electrical system would be risky
without circuit breakers and fuses.
These two invaluable gadgets
warn of any serious malfunction
in the system, and prevent people
from drawing on too much power
at one time.

If something serious goes
wrong, a fuse will blow out or a
circuit breaker will activate. The
fault may lie in an appliance
which is working imperfectly, with
wires that are wearing out, or
with a circuit burdened with
supplying power to too many
appliances at once.

Usually the problem lies in too
many appliances on a single
circuit. You should disconnect the
appliances on a circuit and
replace the blown fuse. If a new
fuse blows out even after
appliaces on the particular circuit
have been disconnected, then the
problem may be in the wiring.
You may have to call in the
services of a professional to fix this
kind of problem.

Provide your house with
special circuits for heavy-duty
appliances such as the range and
washing machine. Spare fuses
should always be on hand in the
vicinity of the fuse box. A
flashlight should be nearby too.
Teach everyone in the house the
right way of putting in a new fuse.
Don't under any circum-
stances replace blown fuses with
pennies, pieces of wire, or
anything other than a new fuse.
Use fuses no larger than fifteen
amps for your household lighting
circuits.

Try to avoid blowing a fuse or
tripping a circuit in the first place
by using electric lights and
appliances with moderation. It's
helpful to learn what appliances
are plugged into which circuit.

One component of the
electrical system that hasn't been
mentioned is electrical outlets, the

connecting point between appliances and circuits. You'll notice that from time to time a plug will not stay in a wall socket securely. This is a clear sign that either the outlet or the plug (or the entire appliance) is defective. Determine what is wrong and fix it.

### Electrical Cords
People sometimes fail to take proper care of the electrical cords attached to the end of appliances. They tie them in knots, suspend them from rails, place them too close to heat, and fling them about like they were indestructible. In fact, cords should be carefully scrutinized along the entire length for breaks in the outer covering and fraying of the inner cable. If a nick is discovered, the whole cord must be replaced. A piece of electrical tape patched around a break does not provide the same safety as a new, defect-free cord.

Electrical cords should not be allowed to knot, as knotting adds to wear and tear. Don't allow cords to get wet. Cords should not be strung alongside of hot spots like radiators or toaster ovens. The heat of such appliances could break open a cord. Above all, cords should be kept in plain view. Putting them underneath rugs and carpeting will subject them to the constant tread of people's steps.

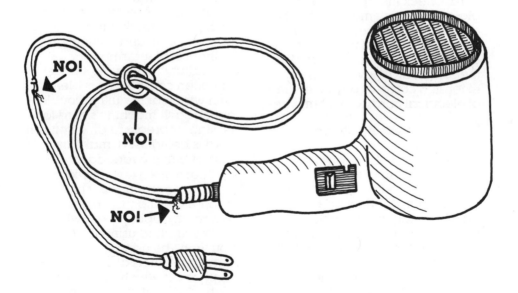

Extension cords are possible fire hazards that deserve special mention. The problem is that extension cords are often the wrong size for the appliance they're connected to. If an apparatus demands more electrical current than an extension cord was designed for, the cord will overheat and a fire will break out. Extension cords are not to function as permanent attachments. They should only be used for short periods of time. If used, they should be of a size and type cord supplied with the appliance. (Use only UL-listed cord sets.)Don't overload an extension cord. The total should not be over 700 watts for a "regular" (No. 18 wire) extension cord. Check the appliance name plate to see how many watts (the electric load) it uses. Total the numbers of watts if more than one appliance is being connected to the same cord. Never tangle up extension cords, or any other kind of electrical cord, with other wires

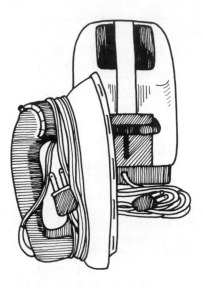

## 7. Malfunctioning Electrical Appliances

Underwriters Laboratories is a national industrial testing laboratory, unaffiliated with manufacturers, which sets standards for, among other products, household appliances. Any appliance you buy should be approved by Underwriters Laboratories, or by another recognized testing outfit.

If you should get a "tingling sensation" when you touch an electrical appliance, don't wait for a severe shock, unplug it at once. If food is spoiling in a refrigerator that was never troublesome before — for any such problems, shut the appliance off and contact an authorized repairman.

Malfunctioning appliances are often caused by problems with wiring or circuitry. Double-check each machine's cord for fraying or breaks. If an extension cord is being used, make certain it is not being overloaded.

There are a few well-known rules for operating any household appliance. The most basic one is to turn an appliance off when you've finished using it, or when you won't be using it for a significant interval of time. Some appliances which are left on all night can heat up enough to ignite surrounding materials.

Larger appliances such as television sets were not designed to be tabletops. Don't set plants or water glasses on television sets or any other large electrical device. If water gets inside, it may cause trouble. Similarly, keep small electric appliances you use in the bathroom, like shavers, hair dryers, and toothbrushes well away from the water.

When operating electric-powered tools, be sure you're not standing in water. Don't operate them out in the rain.

### Take Good Care of Your Appliances

Keep the motors of large appliances and power tools tuned up. Oil and clean them regularly.

Don't leave a hot iron unattended, and never wrap a cord around any appliance that is warm.

Have cooling fans serviced annually. Ventilation fans in the attic should be oiled periodically and electrical connections closely examined.

Provide ample air circulation around your television set, record player and radios.

## Appliances

### Some Common Hazardous Conditions

| Condition | Remedy |
| --- | --- |
| Lights dim when appliance comes on —circuit overloaded. | Change to different circuit, or install new circuit. |
| Lights "flicker." | Tighten loose wire connections in lamp, fixture, switch or circuit. |
| Lights "flicker." | Replace faulty switch. |
| Repeated blowing of fuses or tripping of circuit breakers. | See procedure described on Page 24. |
| Overloaded circuit. | Transfer loads to other circuits. |
| Overloaded circuit. | Install additional circuit. |
| Loose wires (high resistance short). | Tighten all wire connections on the circuit. |
| Short circuit. | Repair or replace faulty appliance, lamp, cord, switch, etc. |
| Appliance cord or plug hot to the touch. | Worn plug or receptacle —replace. Too much load connected to cord. Cord wire too small. |
| "Tingle" shock from appliance. | Disconnect. Have repaired. |
| Broken switch plate. | Replace. |
| Cords or Christmas light string in poor condition. | Replace. Do not attempt to repair. |

Copyright © 1976, National Fire Protection Association — Reproduced With Permission.

## Kitchen Cooking Fires
### Emergency Actions

| Condition | Remedy |
|---|---|
| Small pan fire. | 1. Turn off the heat. |
| | 2. Try smothering the flames by completely covering the pan with a pan cover or cookie sheet. Care is needed to avoid exposing your flesh or clothing in placing the cover on the pan or to upset the pan in your haste. Act quickly, but carefully and deliberately. |
| | 3. Leave the cover or cookie sheet on the pan long enough to allow the fire to smother. Leave it there for several minutes and make sure the stove heat is turned off. |
| | 4. If the pan cover does not smother the fire, or if no cover of proper size is available, use a dry chemical or carbon dioxide extinguisher, but stand 5 or 6 feet away from the pan to avoid spreading the burning material with the stream from the extinguisher. |
| Fire in oven. | 1. Turn off the heat. |
| | 2. Try to smother flames by closing — or keeping closed — the oven door for a few minutes. |
| | 3. If this does not work, open the oven door a crack and use a dry chemical or carbon dioxide extinguisher, closing the door thereafter until flames are obviously out. When opening oven door, open it only enough to see any flames or to see that the fire is out. |
| Drapes or curtains too close to stove. | Use ties to keep curtains away or use different type of window dressing. |
| "Auto exhaust" type of odor in kitchen when cooking or baking. | Call service man or gas utility company to inspect burners — or check air inlet holes at burners for clogging. |
| Smell of gas. | Provide ventilation until remedy is complete. Check to see that burners are off and pilot light is lit. |
| Gas pilot won't light or stay lit. | Check pilot burner for clogging — call service man to adjust pilot. |
| Under surface of kitchen range hood feels greasy. | Clean surface and filter. |

Copyright © 1976, National Fire Protection Association — Reproduced With Permission.

## 8. Cooking In The Kitchen

Fires in the kitchen are most often grease fires. A stove burner or the intense heat of an oven can ignite fat and turn a dish or pan into an inferno.

Kitchen fires usually develop very quickly. The important thing is to act quickly but carefully. Handle the situation calmly. Improper action taken against a grease fire may only lead to spreading the fire to the rest of the house.

### Pan Fires

If a pan catches on fire, do not attempt to carry the pan outside the kitchen. If you try, you're likely to splatter bits of flaming fat onto the floor or curtains. Turn off the heat. Put the pan down on a counter. Put the lid or a cookie sheet over the pan if you can to try to smother the flames. Cutting the flames off from the air should end the fire. Otherwise try pouring baking soda into the pan. **Warning:** never pour water over a grease fire.

### Oven Fires

If a fire develops in the oven, close the oven door. Turn off the heat if it is possible to reach the controls. If the fire doesn't go out, prepare a fire extinguisher for action (one of the dry chemical or carbon dioxide types should always be within easy reach in the kitchen). Open the oven slowly and spray inside.

### Kitchen Fire Safety Tips

Care in preparing food can prevent most kitchen accidents from occurring. Keep countertops generally clean and wiped down. Keep curtains tied back away from the stove and don't hang hot pads or towels over the stove. Keep stove and hood clean and free of grease. Regularly drain excess grease from pans. When roasting something that produces a lot of juices or fats, use a deep enough dish so they don't overflow.

Never turn a burner up to the point where flames are licking at the edge of a pan. Try to avoid broiling food that isn't properly thawed, and never heat frozen foodstuffs at excessive temperatures to make them "cook faster". Stay with frying foods.

Wear tight or short sleeves when you cook. Loose-fitting garments can catch fire.

Don't store things over the stove. People get burned reaching.

## 9. Improper Use of Heating Units

Of the major utilities such as heating and electricity, it is heating which is the cause of most fires. Fires breaking out around heating systems hold special peril because of the considerable amount of combustible fuel held in the storage tanks. Most fires can be avoided by regular inspection and maintenance of the heating system.

Heating fuels like oil and gas, despite their potentially explosive nature, do not themselves constitute a grave danger. The problem arises when leaks develop in fuel containers. Leaks expose the fuel to air, and set up a situation in which uncontrolled combustion can take place. Be watchful for leaks and repair them when they occur.

All heating systems should have adequate clearance around them. A space of eighteen inches should be the minimum distance from the breeching, flues, smoke and stove pipes, or top of furnace or boiler to the ceiling. The intense heat could ignite combustibles placed nearby. Clear away objects within a few feet of the furnace.

This section is broken down into two parts: oil and gas heaters. For more information on wood and coal burning stoves, see the section on Rural Houses in Chapter IV.

## Oil Heating

Safe oil heating largely has to do with regular inspection of the oil storage tank. Watch out for leaks in the storage containers. A defective tank must be replaced

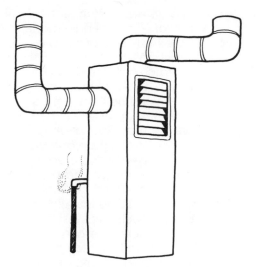

right away. If a tank is leaking, quickly clean up the mess.

Overfilling the storage tank can cause fires too. Overfilling can expose oil to air and give rise to the possibility of ignition. You should know the procedure for determining how much oil is in a tank.

Never keep open flames or heat-producing machinery near an oil tank.

## Gas Heating

Gas heating systems use natural gas and liquified petroleum gas for fuel. The gas is piped to a furnace from a liquified petroleum gas tank, or from a gas meter. All systems should be equipped with standard safety features. If you're not familiar with the parts or working of a gas system, have an expert inspect it.

As with any heating device, proper clearance is a key safeguard. A gas meter, for instance, should have at least 3 feet (a yardstick) of open space between it and any other furnace. Service ducts leading to the furnace should be positioned so that gas escapes into the air in the event of a system malfunction.

A pool of oil on the basement floor is a tell-tale sign of trouble with an oil heater. With gas heaters, watch out for a strong odor of gas as an indication of leakage. If you smell gas from a cellar gas tank, open all cellar windows. It's important to let outside air in to dissipate any gas that has collected. By no means should you use an open flame like a match to inspect the tank. Any machinery in the vicinity powered by electricity must be shut down. You may wish to search for a leak in the fuel container yourself. Sponge soapy water onto the surface of the tank and keep a sharp eye out for bubbles. You may lack confidence in your ability to locate a leak or diagnose equipment failure. In this case an experienced workman should do the job. Don't let the problem sit, in either case. Address it right away.

Liquid petroleum gas is usually not stored in the cellar. Outdoor containers are constructed to hold this type of fuel. Gas containers should be placed far away from other structures. Also, it is a mistake to place liquified petroleum gas tanks below the surface of the ground. This is so because this type of gas is heavier than air, unlike natural gas, which is lighter than air. Leaking liquified petroleum gas can sink far into the ground and may eventually seep into a basement. So the general rule of keeping proper clearance applies here too. A minimum of 3 feet of empty space should stand between a liquified petroleum gas container and any building.

A fire in or around a liquified petroleum gas tank or an outdoor oil tank is a call for swift action. Hose down the affected area. If there is doubt about extinguishing the blaze, call the fire department.

**Space and Kerosene Heaters**
Portable heating units (space heaters) should be operated only in safe areas, never near water. Combustible materials must be kept away from the metal coils, which reach high temperatures. (There should be a guard around the coil.) And they should have tip-over switches to shut the current off if they are knocked over. Portable devices are to be rated by Underwriters Laboratories or Factory Mutual, an insurance company dealing in occupational coverage. Operate them only according to manufacturers' recommendations.

## Heating Equipment
### Some Common Hazardous Conditions

| Condition | Remedy |
| --- | --- |
| Unit poorly designed and without adequate safeguards. | Purchase only tested units labeled by UL, AGA or FM. |
| Combustibles too close. | Keep combustibles well away from heating units. Keep area clean. |
| Odor of gas. | Call utility or plumber. Open windows. Do not use open flames. Use soap solution to look for leaks. |
| Oil fill pipe inside. | Relocate outside and 2 feet from any opening. |
| Oil leaks. | Replace piping and/or tank. |
| No fuel shut-off valve, or valve not handy | Relocate, or install at a readily accessible spot. |
| Gas regulator vent clogged. | Clean out. Install rainshield and insect screen. |
| Wood or cardboard ash box. | Replace with metal container. Keep off wooden floor. |
| Crack in flue or chimney. | Repair or replace. |
| Clogged air filter. | Replace. |
| Holes in flue connector. | Replace with corrosion-resistant pipe. |
| Connector or flue within 6-9 inches of combustibles. | Relocate, or replace with special unit. |
| Heavy tar or soot deposits inside chimney or flue. | Have cleaned. |
| Floor furnace register in passageway. | Relocate, or provide guards. |
| Heater located in hall or near door, or stairway. | Relocate out of exit paths and well well away from doors and stairways. |
| Kerosene or gasoline used to "freshen" stove fire. | An extremely dangerous practice to be absolutely prohibited. |
| Storage in heater closet. | Remove all combustibles. |
| Heater of furnace in garage. | Flame box should be 18 inches or more above the floor. |

Copyright © 1976, National Fire Protection Association — Reproduced With Permission.

## 10. Uncontrolled Open Flames

Fireplaces, lanterns, and grills can provide heat, light, food, and the invaluable commodity of a comfortable atmosphere. They also produce open flames and so pose an obvious fire hazard. Open flames must be kept under control.

The space around a fireplace should be free of papers or firewood. Learn the proper procedure for igniting a blaze. (Never use highly combustible liquids like gasoline or benzene.) Have the proper implements for stoking a fire next to the hearth. After all, its risky to turn over logs with a piece of lumber. A tight fitting screen should be in front of the fireplace to protect the room from flying embers. Regularly clean out the interior of the chimney.

When burning leaves and rubbish outdoors, use a suitable incinerator, keep water handy and never leave a fire unattended.

With barbecue grills, again, don't use gasoline or kerosene as a fuel. Stick with the lighter fluid sold in supermarkets, but don't pour excessive amounts over the charcoal. Take heed of the wind's direction in cooking, and beware of shooting flames caused by dripping fat.

Two more points should be made about barbecues. Never place a grill on a porch, stairway, or too close to a dwelling. Keep it

well away from shrubs, trees and wooden fences. And don't leave your children unsupervised while they're roasting marshmallows or cooking popcorn over a flame. This applies to both fires inside and outside the house.

Lanterns and candles really should not be open flames at all. Equip them with a glass sheathing acting as a barrier between flame and exterior combustibles. If lanterns are attached to a wall, fastenings must be secure. After each use a lantern or candle holder should be cleaned out. A candle never should be placed on a table unless it is supported by a holder.

Fondues are an increasingly popular dish, and a few precautions should keep fondues relatively risk-free. Use an alcohol lamp to keep the dip in the fondue basin warm. Place the lamp only on a solid wood or plastic table. (Don't put a tablecloth under the lamp under the burner.) To prevent spillage keep the food basin less than half full at all times.

## 11. Thawing Frozen Water Pipes

People have been known to engage in the most reckless behavior to thaw out water pipes frozen by a winter cold snap. There are many documented cases of impatient residents using blowtorches against their own plumbing in a careless, hasty attempt to melt the ice. Needless to say, blowtorches or torches employed in this manner have started many destructive fires.

Much of the risk in thawing water pipes can be eliminated by using electrical heating units that have been specially designed for this very problem. These devices negate the fire hazard of working in the cramped enclosed areas, such as within basement walls or underneath floors, where piping is often housed.

The expense and hassle of thawing out pipes can be avoided altogether by keeping an ear fixed to the weather report. If you receive advance warning of frigid weather, turn off the water in your home, or allow the faucets to drip slightly.

The most important point is this: if you must take a torch or other unsafe heating source to a pipe, after you're done closely inspect the surrounding area for burn marks. It may take hours for a few square inches of smoldering wood to turn into a raging fire. So check carefully.

## FIRE PROTECTION CHECKLIST

The checklist that follows is not intended as a comprehensive test of fire safety. It will give you a general idea of how fire-safe your home is, and may point out some of the hazards you've overlooked. A "Yes" answer means that you are practicing fire safety for the particular item. A "No" answer mean that you have a potentially dangerous condition to correct.

The questions are grouped into major categories of fire hazards. At the end of the checklist, tally up your answers to see how your home rates in fire safety.

## Smoking Habits And Care Of Matches

| Yes | No | |
|-----|-----|---|
| | | Do you keep matches and lighters out of the reach of small children? |
| | | Are there plenty of good-sized noncombustible ashtrays in areas where smoking is permitted? |
| | | Are tobacco ashes disposed of in their own separate containers? |
| | | Have you made it a house rule never to smoke in bed or when dozing on the sofa? |
| | | Has every member of your household been instructed never to use matches or candles to light the way in the attic, closets, or basement? |
| | | After having company over, do you check the floor and your furniture for smoldering butts or ashes? |

(continued on next page)

## Central Heating Hazards

| Yes | No | |
|-----|-----|---|
| | | Do you have the entire heating system (including burner, flue pipes, chimney, and vents) inspected, cleaned, and repaired by a professional service each year? |
| | | Is your heating system a type listed by Underwriters Laboratories or, if gas-fired, by the American Gas Association Laboratories? |
| | | Have you eliminated all flue pipes and vent connectors passing through closets, floors, ceilings, and attic? |
| | | Are the ceiling, walls, and partitions around all parts of the heating system protected by noncombustible material adequately separated from the source of heat? |
| | | Is your inside basement door at the head of the stairs well fitted and kept closed at night? |
| | | Are all room heaters placed on a steady, level foundation? |
| | | Do room heaters have ample air circulation around them? |
| | | Are room heaters located apart from curtains, furniture, and clothing? |

## Housekeeping Hazards

| Yes | No | |
|-----|-----|---|
| | | Do you always place rags covered with oil or paint in metal cans after you use them? |
| | | If you store paints, solvents, waxes, etc., are they in tightly closed cans, away from heat, flames and sparks? |
| | | Have you forbidden household members to use gasoline, benzene, or any other inflammable fluid for cleaning clothes, furnishings or floors? |
| | | Is it a rule in your home never to start a fire in a fireplace, stove or furnace with an inflammable liquid? |
| | | Do you keep ashes from a furnace in a metal container away from combustible materials, and dispose of them frequently? |
| | | Is your kitchen stove, oven, and rotisserie kept clean of grease? |
| | | If you have a woodworking shop, do you clean up scrap wood and sawdust after each job? |

(continued on next page)

## Electrical Hazards

| Yes | No | |
|-----|-----|---|
| | | Are all of your electrical appliances, small and large, listed by Underwriters Laboratories? |
| | | Does every room have enough electrical outlets to avoid the need for multiple attachment plugs? |
| | | Does your home have special circuits for heavy-duty appliances such as ranges and washing machines? |
| | | Do you use fuses no larger than 15 amps for your household lighting circuits? |
| | | If you use extension cords, are they in good condition, and are they out in the open rather than under rugs, over hooks, or through door openings and partitions? |
| | | Is there ample air circulation around your television set and radios? |
| | | Do your cooking appliances and electric iron have heat-limit controls? |
| | | Are the motors of your large appliances and power tools cleaned and oiled regularly? |
| | | Do you refrain from wrapping cords around a hot appliance? |
| | | Do you ever use more than one high-wattage appliance on an outlet at a time? |

(continued on next page)

## Basement, Garage and Yard Hazards

| Yes | No | |
|-----|-----|-----|
| | | Are lawn mowers and all gas-powered machines stored with empty tanks? |
| | | Is charcoal lighter fluid stored in a sealed metal container? |
| | | Do you cover up unused electrical outlets? |
| | | Are electrical appliances unplugged when not in use? |
| | | When you store gasoline do you keep it in an approved metal safety can? |
| | | Do you keep gasoline for power tools in safety-type metal cans, and do you refuel out-of-doors? |
| | | Is your yard, or any adjacent vacant property, kept clear of dry leaves and combustible litter? |
| | | When leaves and rubbish are burned outdoors, do you use a suitable incinerator, keep water handy, and never leave fire unattended? |
| | | When making an outdoor fire (for barbecueing or for burning trash), do you always make sure the fire is set well away from shrubs, trees, and wooden fences? |

## Holiday Hazards

| Yes | No | |
|-----|-----|-----|
| | | Do you put fire-resistant decorations on your walls during the days surrounding holidays? |
| | | Is your Christmas tree kept well watered? |
| | | Is the Christmas tree treated with flame-retardant spray? |
| | | Do you keep children from playing under the tree? |
| | | Do you extinguish Christmas tree lights when you leave the room? |
| | | Do you use the correct size electrical cords with holiday lighting? |

Chapter II

# THERE'S NO PLACE LIKE HOME
## What To Check When Building Buying or Renting

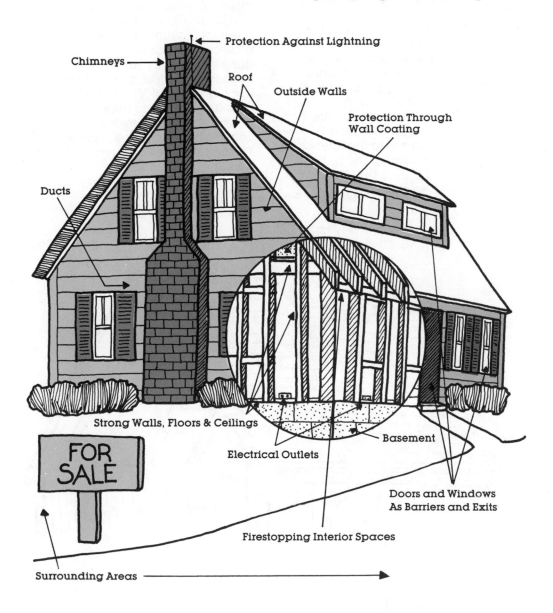

Protection Against Lightning

Chimneys

Roof

Outside Walls

Protection Through
Wall Coating

Ducts

Strong Walls, Floors & Ceilings

Electrical Outlets

Basement

FOR SALE

Doors and Windows
As Barriers and Exits

Firestopping Interior Spaces

Surrounding Areas

## Go Looking For Trouble

Let's take a tour of the typical home section-by-section to determine what your concerns should be:

## Have A Secure Roof And Outside Walls

A solidly constructed roof of fire-resistant material is a great advantage in fire protection. A superior roof will fend off sparks from blazes taking place outside the house. A good roof is invaluable too in case of a fire inside the house. It will tend to contain the flames to the interior. By helping to firm up the rest of the building, it may buy time for residents to escape.

Roofs should consist of fire-resistant shingles or roofing made of materials like asphalt, slate, metal, or cement. Ordinary wood shingles should never be used for outside walls or as roofing material. While they are very quaint and rustic-looking, they are next to worthless as a protection against fire. Wood shingles are very susceptible to ignition by sparks, particularly upon drying out. The fire resistance of wood shingles can be enhanced somewhat by coating them with fire retardants. Treated shingles are available for building a new roof.

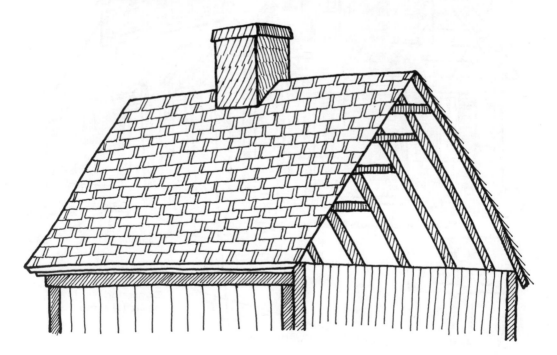

## Typical Prepared Roof Coverings*

| Description | Minimum Incline In. to Ft. | CLASS A |
|---|---|---|
| **Asphalt-Asbestos Felt Sheet Coverings** | Not exceeding 12 | Factory-assembled sheets of 4-ply asphalt and asbestos material. |
| **Asphalt-Asbestos Felt Shingle Coverings** | Exceeding 4 | |
| **Organic Felt (previously referred to as rag felt) Sheet Coverings** | Exceeding 4 | |
| **Organic Felt (previously referred to as rag felt) Shingle Covering, with Special Coating** | Sufficient to permit drainage. | Grit surfaced, two or more thicknesses. |
| **Organic Felt (previously referred to as rag felt) Shingle Coverings** | Sufficient to permit drainage. | Grit surfaced, two or more thicknesses. |
| **Asphalt Glass Fiber Mat Shingle Coverings** | Sufficient to permit drainage. | Grit surfaced, two or more thicknesses. |
| **Asphalt Glass Mat Sheet Covering** | Sufficient to permit drainage. | |
| **Fire-Retardant Treated Red Cedar Wood Shingles and Shakes** | Sufficient to permit drainage. | |

*Prepared roof coverings are classified as applied over square-edge wood sheathing of 1-in. nominal thickness, or the equivalent, unless otherwise specified. See Built-up Roof Coverings (page 48). Laid in accordance with instruction sheets accompanying package. Limited to decks capable of receiving and retaining nails.

Where organic felt is indicated, asbestos felt of equivalent weight can be substituted.

By end lap is meant the overlapping length of the two units, one placed over the other. Head lap in shingle-type roofs is the distance a shingle in any course overlaps a shingle in the second course below it. However, with shingles laid by the Dutch-lap method, where no shingle overlaps a shingle in the second course below, the head lap is taken as the distance a shingle overlaps one in the next course below.

Prepared roofings are labeled by Underwriters Laboratories which indicate the classification when applied in accordance with directions for application included in packages.

(continued on next page)

**Typical Prepared Roof Coverings\*** (continued from previous page)

| Description | Minimum Incline In. to Ft. | Class B | Class C |
|---|---|---|---|
| **Asphalt-Asbestos Felt Sheet Coverings** | Not exceeding 12 | Factory-assembled sheets of 3-ply asphalt and asbestos material or sheet coverings of single thickness with a grit surface. | Single thickness, smooth surfaced. |
| **Asphalt-Asbestos Felt Shingle Coverings** | Exceeding 4 | Asphalt-asbestos felt, grit surfaced. | |
| **Organic Felt (previously referred to as rag felt) Sheet Coverings** | Exceeding 4 | | Sheet coverings of asphalt organic felt, either grit surfaced or aluminum surfaced. |
| **Organic Felt (previously referred to as rag felt) Shingle Covering, with Special Coating** | Sufficient to permit drainage. | Grit surfaced, two or more thicknesses. | |
| **Organic Felt (previously referred to as rag felt) Shingle Coverings** | Sufficient to permit drainage. | Grit surfaced, two or more thicknesses. | Grit surfaced shingles, one or more thicknesses. |
| **Asphalt Glass Fiber Mat Shingle Coverings** | Sufficient to permit drainage. | Grit surfaced, one or more thicknesses. | Grit surfaced shingles, one or more thicknesses. |
| **Asphalt Glass Mat Sheet Covering** | Sufficient to permit drainage. | | Grit surfaced. |
| **Fire-Retardant Treated Red Cedar Wood Shingles and Shakes** | Sufficient to permit drainage. | | Treated shingles or shakes, one or more thicknesses; shakes require at least one layer of Type 15 felt underlayment. |

(continued on next page)

| Description | Minimum Incline In. to Ft. | CLASS A |
|---|---|---|
| **Brick** **Concrete** **Tile** **Slate** | | Brick, 2¼ in. thick. Reinforced portland cement, 1 in. thick. Concrete or clay floor or deck tile, 1 in. thick. Flat or French-type clay or concrete tile, ⅜ in. thick with 1½ in. or more end lap and head lock, spacing body of tile ½ in. or more above roof sheathing, with underlay of one layer of Type 15 asphalt-saturated asbestos felt or one layer of Type 30 or two layers of Type 15 asphalt-saturated organic felt. Clay or concrete roof tile, Spanish or Mission pattern, ⁷⁄₁₆ in. thick, 3-in. end lap, same underlay as above. Slate, ³⁄₁₆ in. thick, laid American method. |
| **Metal Roofing** | 12 | Sheet roofing of 16-oz copper or of 30-gage steel or iron protected against corrosion. Limited to noncombustible roof decks or noncombustible roof supports when no separate roof deck is provided. |
| **Cement-Asbestos Shingles** | Exceeding 4 | Laid to provide two or more thicknesses over one layer of Type 15 asphalt-saturated asbestos felt. |

(continued on next page)

## Typical Prepared Roof Coverings* (continued from previous page)

| Description | Minimum Incline In. to Ft. | CLASS B | CLASS C |
|---|---|---|---|
| **Brick Concrete Tile Slate** | | | |
| **Metal Roofing** | 12 | Sheet roofing of 16-oz copper or of 30-gage steel or iron protected against corrosion or shingle-pattern roofings with underlay of one layer of Type 15 saturated asbestos felt, or one layer of Type 30 or two layers of Type 15 asphalt-saturated organic felt. | Sheet roofing of 16-oz copper or of 30-gage steel or iron protected against corrosion or shingle-pattern roofings, either without underlay or with underlay of rosin-sized paper. |
| | | | Zinc sheets or shingle roofings with an underlay of one layer of Type 30 or two layers of Type 15 asphalt-saturated organic felt or one layer of 14 lbs unsaturated or one layer of Type 15 asphalt-saturated asbsestos felt. |
| **Cement-Asbestos Shingles** | Exceeding 4 | Laid to provide one or more thicknesses over one layer of Type 15 asphalt-saturated asbestos felt. | |

Copyright © 1976, National Fire Protection Association — Reproduced With Permission.

## Built-up Roof Coverings*

| Description | Minimum Incline In. to Ft. | CLASS A |
|---|---|---|
| **Asphalt organic felt, bonded with asphalt and surfaced with 400 lbs of roofing gravel or crushed stone, or 300 lbs of crushed slag per 100 sq ft of roof surface, on coating of hot mopping asphalt** | 3 | 4 (plain) or 5 (perforated) layers of Type 15 felt.<br><br>1 layer of Type 30 felt and 2 layers of Type 15 felt.<br><br>1 layer of Type 15 felt and 2 layers of Type 15 or 30 cap or base sheets.<br><br>3 layers of Type 15 or 30 cap or base sheets.<br><br>3 layers of Type 15 felt. Limited to noncombustible decks. |
| **Tar-asbestos felt or organic felt bonded with tar and surfaced with 400 lbs of roofing gravel or crushed stone, or 300 lbs of crushed slag per 100 sq ft of roof surface on a coating of hot mopping tar** | 3 | 4 layers of 14-lb asbestos felt or Type 15 organic felt.<br><br>3 layers of 14-lb asbestos felt or Type 15 organic felt. |
| **Steep tar organic felt** | 5 | 4 layers of Type 15 tar-saturated organic felt, bonded with steep coal-tar pitch, surfaced with 275 lbs of ⅝-in. crushed slag per 100 sq ft of roof surface on steep coal-tar pitch. |
| **Asphalt organic felt, plain or perforated, bonded and surfaced with a cold application coating** | 12 | |

*Built-up roof coverings are classified as applied over square-edge wood sheathing of 1-in. nominal thickness, or the equivalent, unless otherwise specified.

From the stand point of relative effectiveness of the different types of wood roof sheathing, the tongue-and-groove boards and ¾-in. moisture-resistant plywood give better results in the brand and flame tests than square-edge sheathing with boards spaced about ¼-in. apart. For classifications based on square-edge sheathing, tongue-and-groove or plywood sheathing can be substituted. Square-edge sheathing boards should be butted together as closely as possible. Reference to ¼-in. spacing is to indicate fire test procedure intended to simulate actual conditions after shrinkage of boards due to age or other reasons.

The minimum weight of cementing material between separate layers of felt is considered to be 25 lbs per 100 sq ft of roof surface.

Types 15 and 30 felts are defined as saturated felts weighing a minimum of 14 lbs and 28 lbs per 100 sq ft of the finished materials, respectively. Where saturated felts are referred to by weight, the weight is minimum and is expressed in pounds per 100 sq ft of the finished material.

Materials intended for built-up roof coverings are labeled by Underwriters Laboratories. The classifications indicated are of generally accepted combinations.

(continued on next page)

## Built-up Roof Coverings* (continued from previous page)

| Description | Minimum Incline In. to Ft. | Class B | CLASS C |
|---|---|---|---|
| **Asphalt organic felt, bonded with asphalt and surfaced with 400 lbs of roofing gravel or crushed stone, or 300 lbs of crushed slag per 100 sq ft of roof surface, on coating of hot mopping asphalt** | 3 | 4 layers of perforated Type 15 felt. <br> 3 layers of Type 15 felt. <br> 2 layers of Type 15 or 30 cap or base sheets. | |
| **Tar-asbestos felt or organic felt bonded with tar and surfaced with 400 lbs of roofing gravel or crushed stone, or 300 lbs of crushed slag per 100 sq ft of roof surface on a coating of hot mopping tar** | 3 | 3 layers of 14-lb asbestos felt or Type 15 organic felt. | |
| **Steep tar organic felt** | 5 | 4 layers of Type 15 tar-saturated organic felt, bonded with steep coal-tar pitch, surfaced with 275 lbs. of ⅝ in. crushed slag per 100 sq.ft. of roof surface on steep coal-tar pitch. | |
| **Asphalt organic felt, plain or perforated, bonded and surfaced with a cold application coating** | 12 | | 3 layers of Type 15 felt. <br> 1 layer of Type 30 felt and 1 layer of Type 15 felt. <br> 2 layers of Type 15 or 30 cap or base sheets. <br> 2 layers of Type 15 felt and 1 layer of Type 15 or 30 cap or base sheets. |

Copyright © 1976, National Fire Protection Association — Reproduced With Permission.

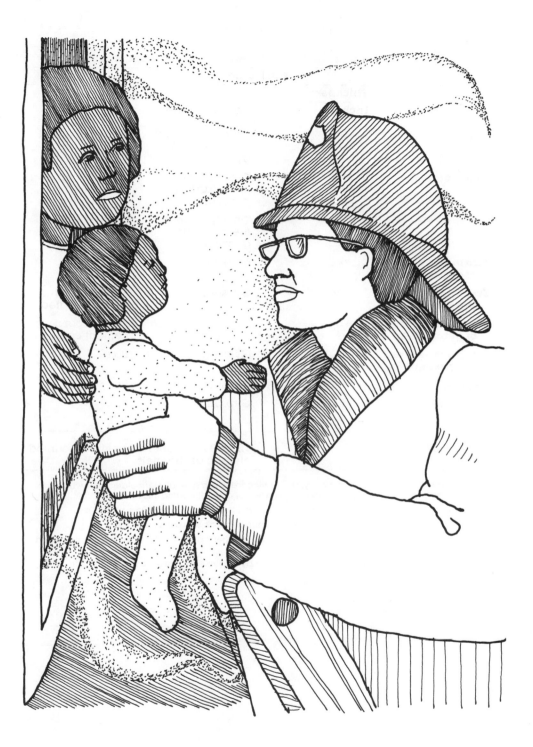

## Wood Joist Floors with Wallboard Ceilings (Combustible)

| Type | Wallboard Thickness Inches | Core Materials | Type |
|------|------|------|------|
| **Gypsum** | ⅝ | Type "X"* special fire-retardant gypsum | Cement-coated wire |
| **Gypsum** | ½ | Type "X"* special fire-retardant gypsum | Cement-coated wire |
| **Gypsum** | ⅜ | Type "X"* special fire-retardant gypsum | Cement-coated wire |
| **Two thicknesses of gypsum** | ½ + ½ | Gypsum | Box wire |
| **Two thicknesses of gypsum** | ⅜ + ½ | Gypsum | Plasterboard cement-coated |
| **Two thicknesses of gypsum** | ½ + ⅜ | Gypsum | Plasterboard cement-coated |
| **Two thicknesses of gypsum‡** | ⅜ + ⅜ | Gypsum | Box |
| **Gypsum** | ½ | Gypsum | Box |
| **Gypsum§** | ⅜ | Gypsum | Box |
| **None** | — | — | — |
| **Acoustical Tile** | ⅝ | 12- by 12-in. mineral fiber tiles mounted on special channels | |

\* Type X gypsum wallboard designates gypsum wallboard with a specially formulated core which provides greater fire resistance than regular gypsum wallboard of the same thickness.

† 1-in. hexagonal mesh 20-gage wire fabric between wallboards nailed with 8d nails 8 in. o.c.

‡ NBS test on floor 4½ by 9 ft joints of wallboard staggered, but no tape and joint finisher.

§ NBS test: bottom of ceiling covered with 14 lb asbestos paper applied with paperhanger's paste and casein paint.

‖ NBS test on 2 specimens of open-joist floors, each 4½ by 9 ft; fire endurance 15 min and 12 min.

## Wood Joist Floors with Wallboard Ceilings (Combustible)

| Nails | | | | Fire Resistance Rating | |
|---|---|---|---|---|---|
| Size | Gage | Length Inches | Spacing Inches | Hr | Min |
| 6d | 13 | 1⅞ | 6 | 1 | — |
| 5d | 13½ | 1⅝ | 6 | — | 45 |
| 4d | 14 | 1⅜ | 6 | — | 30 |
| 5d(1)<br>6d(2) | 14<br>10¼ | 1¾<br>2½ | 18<br>6 | 1† | — |
| 1½ in<br>6d | 13<br>13 | 1½<br>1⅞ | 7<br>6 | — | 40 |
| 1½ in<br>6d | 13<br>13 | 1½<br>1⅞ | 7<br>6 | — | 35 |
| 4½d(1)<br>4½d(2) | —<br>— | 1½<br>1½ | 6<br>6 | — | 35 |
| 4½d | 15 | 1½ | 6 | — | 25 |
| 4½d | 15 | 1½ | 6 | — | 25 |
| — | — | — | — | — | 14 |
| | | | | 1 | — |

Copyright © 1976, National Fire Protection Association — Reproduced With Permission.

## Protection Against Lightning

Your locality may or may not fall prey to lightning storms. If it does, and your home lacks a lightning protection system, you may want to consult a contractor or your landlord about obtaining one.

## Strong Walls, Floors, And Ceilings

The ability of interior walls, floors, and ceilings to withstand fire is important because they may turn out to be the only barrier between you and a fire as you

### Wood Stud Walls and Partitions (Combustible)
(Bearing and nonbearing 2- by 4-in studs spaced 16 in. on centers, fire-stopped)

| Material | Fire-Resistance Rating | | | |
| --- | --- | --- | --- | --- |
| | Partition hollow | | Partition filled with mineral wool† | |
| | Hr | Min | Hr | Min |
| **Plasterless Types of Construction** | | | | |
| The following are applied to both sides of studs: | | | | |
| Sheathing boards (tongue-and-groove) ¾ in. thick. | — | 20 | — | 35 |
| Gypsum wallboard, ⅜ in. thick | — | 25 | — | — |
| Gypsum wallboard, ½ in. thick (nonload-bearing only for mineral wool filled) | — | 40 | 1 | — |
| Gypsum wallboard, ⅜ in. thick, in two layers each face | 1 | — | — | — |
| Gypsum wallboard, ½ in. thick, in two layers each face | 1 | 30 | — | — |
| Gypsum wallboard, ½ in. thick, Type X*, one layer each surface | — | 45 | — | — |
| Gypsum wallboard, ⅝ in. thick, Type X*, one layer each face | 1 | — | — | — |
| Gypsum wallboard, ⅝ in. thick, Type X*, on fire-retardant wood fiberboard, ½ in. thick | 1 | — | — | — |
| Fir plywood, ¼ in. thick | — | 10 | — | — |
| Fir plywood, ⅜ in. thick | — | 15 | — | — |
| Fir plywood, ½ in. thick | — | 20 | — | — |
| Fir plywood, ⅝ in. thick | — | 25 | — | — |
| Cement-asbestos board, ³⁄₁₆ in. thick | — | 10 | — | 40 |
| Cement-asbestos board, ³⁄₁₆ in. thick, on gypsum wallboard ⅜ in. thick | 1 | — | — | — |
| Cement-asbestos board, ³⁄₁₆ in. thick, on gypsum wallboard ½ in. thick | 1 | 25 | — | — |

* Type X gypsum wallboard has a special core providing greater fire resistance.

† Mineral wool fill requires some degree of anchorage so as to be held in place after partition facing has been burned away.

(continued on next page)

prepare to make an escape. You should ascertain their thickness and the degree of fire-resistance of their component materials, the latter by looking up the findings of the testing laboratories.

Walls, floors, and ceilings that surround bedrooms in private houses, or individual dwelling units in apartments, should be relatively thick and fireproof. The same might be said of the constructs around fire-prone areas like the kitchen. Your intent should be to contain a fire to its initial area, and shield the rooms

**Wood Stud Walls and Partitions (Combustible)** (continued from previous page)
(Bearing and nonbearing 2- by 4-in studs spaced 16 in. on centers, fire-stopped)

| Material | Fire-Resistance Rating | | | |
| --- | --- | --- | --- | --- |
| | Partition | | Partition filled with | |
| | Hr | Min | Hr | Min |
| **Plaster and Lath Construction** | | | | |
| The following are applied to both sides of studs: | | | | |
| Gypsum-sand plaster, 1:2, 1:3, ½ in. thick. on wood lath | — | 30 | 1 | — |
| Lime-sand plaster, 1:5, 1:7.5, ½ in. thick on wood lath | — | 30 | — | 45 |
| Gypsum-sand plaster, 1:2, 1:2, ½ in. thick on ⅜ in. perforated gypsum lath | 1 | — | — | — |
| Gypsum-sand plaster, 1:2, 1:2, ¾ in. thick on ⅜ in. metal lath | 1 | — | 1 | 30 |
| Gypsum lath ½ in. thick, Type X*, and ⅛ in. gypsum-sand plaster, each face, | 1 | | | |
| Gypsum wallboard, ½ in. thick, Type X* and 1/16 gypsum plaster each face | 1 | — | — | — |
| Portland cement-lime-plaster, 1:1/30:2, 1:1/30:3 and asbestos-fiber plaster, ⅞ in. thick on metal lath. | 1 | — | — | — |
| Gypsum-vermiculite, or perlite plaster, 100 lb gypsum to 2½ cu ft aggregate, ½ in. thick on ⅜ in. perforated gypsum lath | 1 | — | — | — |
| Gypsum-perlite plaster, 1:2, ¾ in. thick on metal lath | 1 | — | — | — |
| **Exterior Bearing Wall** | | | | |
| Outside: Cement-asbestos shingles, 5/32 in. thick, on layer of asbestos felt on wood sheathing, ¾ in. thick, on wood studs. Inside: Cement-asbestos facing, ⅛ in. thick on fiberboard, 7/16 in. thick | — | 30 | — | — |
| Outside: Gypsum sheathing, ½ in. thick. Inside: 1:2 gypsum-sand plaster, ½ in. thick on ⅜ in. perforated gypsum lath | 1 | 30 | — | — |

Copyright © 1976, National Fire Protection Association — Reproduced With Permission.

where people are likely to be.

There are several ways of enhancing the strength of pre-existing constructs. Two methods are firestopping and layering the walls with protective finish. Descriptions of these two techniques follow:

### Firestopping Interior Spaces

"Firestopping" is a term most people don't understand, but it's really a fairly easy concept to grasp. Many walls and floor-ceilings are hollow to a large degree. Smoke, flames, and noxious gases can travel rapidly up the space between enclosed walls and floor-ceilings. The hollows may be several floors in length or height, which means that, for example, a small fire starting in the cellar could shoot

deadly gases up to a third-floor bedroom in a few seconds. Firestopping means the placement of barriers, often wooden boards, in the hollow spaces to slow the spread of fire and gas.

In many homes, and even apartments, open spaces can be seen from the basement or down through the attic. These can be blocked with two-by-fours nailed into the openings or by filling the space with bricks and mortar.

The most insidious quality of fire is its ability to spread with incredible rapidity. Firestopping can greatly slow down a fire. If you're building a new home or refurbishing an old one, insist on firestopping, an inexpensive way to greatly improved home safety.

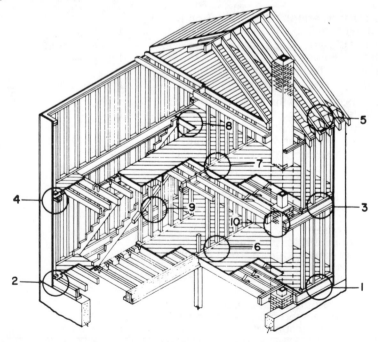

**Example of wood platform frame construction showing points to be firestopped.**

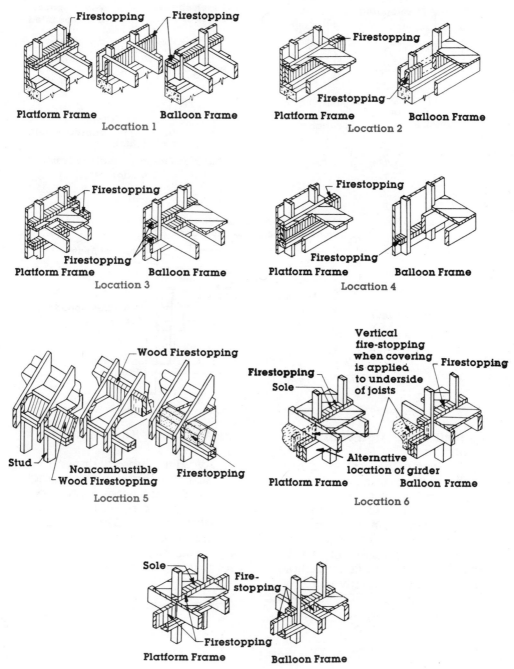

**Firestopping** **Firestopping**

Platform Frame     Balloon Frame
Location 1

**Firestopping**

**Firestopping**

Platform Frame     Balloon Frame
Location 2

**Firestopping**

**Firestopping**

Platform Frame     Balloon Frame
Location 3

**Firestopping**

**Firestopping**

Platform Frame     Balloon Frame
Location 4

Wood Firestopping

Stud

Noncombustible
Wood Firestopping     Firestopping
Location 5

Vertical
fire-stopping
when covering
is applied
to underside
of joists     Firestopping

Firestopping
Sole

Alternative
location of girder
Platform Frame     Balloon Frame
Location 6

Sole

Fire-
stopping

Firestopping

Platform Frame     Balloon Frame
Location 7

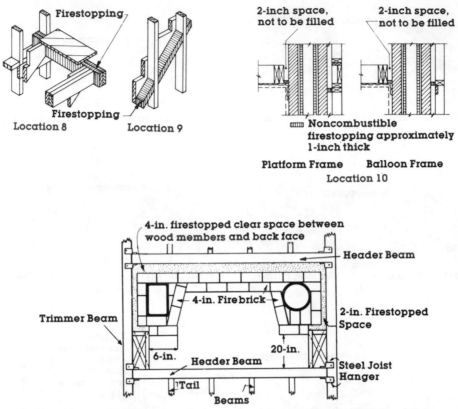

Firestopping

Firestopping

Location 8          Location 9

2-inch space, not to be filled

2-inch space, not to be filled

Noncombustible firestopping approximately 1-inch thick

Platform Frame          Balloon Frame

Location 10

4-in. firestopped clear space between wood members and back face

Header Beam

4-in. Firebrick

2-in. Firestopped Space

Trimmer Beam

6-in.          Header Beam          20-in.

Steel Joist Hanger

Tail          Beams

Copyright © 1976, National Fire Protection Association — Reproduced With Permission.

## Protection Through Wall Coatings

The use of the right kinds of wall coverings can significantly enhance the fire-protective properties of walls and floor-ceilings. Examples of effective interior finish materials are gypsum and plaster. The idea is to use compact, noncombustible material that won't give off choking gases if it does start to burn.

Avoid the types of drywall construction that can easily catch on fire. Certain types of plywood, fireboard, and paperboard fall into this category. The possibility of ignition can be reduced also by coating surfaces with fire-retardant paints.

## Ducts

The section on firestopping described the danger of having empty space in the interiors of walls. Hollows give fire gases space into which they can rapidly expand. Ducts are vertical channels that were built into a house deliberately for a specific purpose. A dumbwaiter is a well-known example. Ducts such as laundry chutes should have tight-fitting covers at their openings.

The covers will keep fire from spreading up and down walls. Kitchen exhaust ducts should lead directly to the outside of the house. In building a house or apartment, ducts should be dispensed with if possible, because they can act as flues for the spread of heat, smoke, gases, and flames.

## Doors And Windows As Barriers And Exits

Doors serve the essential function of blocking a fire's access to large interior spaces. Unfortunately, a dangerous trend in modern home design is the elimination of doors, or the installation of flimsy hollow doors. This constitutes a serious error in fire prevention design. Doors to the basement, the attic, and bedrooms are

especially vital. Indeed, if it is possible to close off a hallway leading to your home's bedrooms by adding an additional door, by all means do so.

Doors should be constructed of densely packed, fire-resistant material. Many doors are solid enough but make a poor fit for the doorway. Doors should be tight-fitting to block the seepage of smoke and gases. House members should be able to open every door in the house from either side of the door. Remember, make a habit of closing your bedroom door before going to sleep.

In a fire, windows function as one avenue of escape. Be certain they can be used for this purpose. All windows should be big enough for a large person to fit through. They should be close enough to the floor that small children can readily get onto the sill. Keep a stepladder next to escape windows high off the baseboards. The area around the windows must be kept clear.

Never nail a window down. Watch out for windows that may

be painted shut. Storm windows or screens should be easy to lift; if they are not, replace them. And instruct children in their use.

## Chimneys

One of the most overlooked fire hazards of today, and also one of the most misunderstood, involves the fireplace, chimney, or woodburning stove.

Becuse of the high cost of oil and electricity, fireplaces and woodburning stoves have become very popular again in the United States and Canada. The unfortunate thing is that most people don't seem to realize that chimneys need regular cleaning when wood is burned, particularly when it is burned on a daily basis during the heating season.

Fire departments throughout the country can attest that recently a major source of fires have come about because of poorly maintained chimneys and fireplaces.

The ideal approach to this problem is to contact a professional chimney sweep to inspect and clear a chimney on a

regular basis. This is mandatory in Switzerland and they have few fires. It's better to have it done in the spring or summer than in the fall and winter.

The fire hazard in chimneys is this: dustlike carbon deposits called creosote collect on the inside of a chimney flue, and impair the fireplace draft. Many fireplace and wood stove manufacturers emphasize the fact that a carbon buildup may not only lessen the efficiency of the unit, but result in a fire because of its extreme flammability. Creosote burns with an intense flame that can melt the mortar inside a chimney and spread the conflagration to upper floors.

Flaming balls of debris can be lifted out of the chimney onto the roof, lawn, or even a neighbor's house. Smoke can back up into the home as well.

Professional associations recommend that chimneys be inspected and cleaned once a year. When a wood burning stove is used regularly, its chimney should be checked every six months. The chimney mortar should be inspected too. Mortar should be tightly packed, never cracked. Unused flue openings should be repaired with solid masonry and periodically checked.

## Construction of Masonry Chimneys

| | Thickness in Inches (Minimum) Walls | Lining | Termination Above Roof, Feet |
|---|---|---|---|
| **Residential Type Appliances** | | | |
| Solid masonry units or reinforced concrete with fire clay lining extending 8 in. below connector or above throat of fireplace | 4 | ⅝ | 3 — 2 ft. from any portion of building within 10 ft |
| Rubble stone masonry with similiar lining | 12 | ⅝ | |
| **Low Heat Appliances** | | | |
| Solid masonry units or reinforced concrete with fire clay lining extending 8 in. below connector or above throat of fireplace | 8 | ⅝ | 3 — 2 ft from any portion of building within 10 ft |
| Rubble stone masonry with similar lining | 12 | ⅝ | |
| **Medium Heat Appliances** | | | |
| Solid masonry units or reinforced concrete with fire brick lining laid in fire clay mortar, lining extending 2 ft below and 25 ft above connector entrance to chimney | 8 | 4½ | 10 ft higher than any portion of building within 25 ft |
| Rubble stone masonry with similar lining | 12 | 4½ | |
| **High Heat Appliances** | | | |
| Double Walls: | | | |
| Outer Wall of solid masonry units or reinforced | 8 | | 20 ft higher than any portion of building within 50 ft |
| Air space | 2 | | |
| Inner wall, total thickness 8 in., of solid masonry units or reinforced concrete with fire clay brick lining laid in fire clay mortar | 3½ | 4½ | |
| **Incinerators** | | | |
| Residential Type — same as for low heat appliances | | | |
| Chute Fed | | | |
| Combined hearth and grate area 7 sq ft or less | | | |
| Clay or shale brick with fire brick lining for 10 ft above roof of combustion chamber | 4 | 4½ | Same as for low heat appliances |
| More than 10 ft above roof of combustion chamber with fire clay flue liner | 8 | ⅝ | |
| Combined hearth and grate area over 7 sq ft | | | 10 ft higher than any portion of roof within 25 ft |
| Clay or shale brick with fire brick lining for 40 ft above roof of combustion chamber | 4 | 4½ | |
| More than 40 ft above roof of combustion chamber with fire clay flue liner | 8 | ⅝ | |
| Commercial or Industirial Types | | | Same as for chute fed |
| Clay or shale brick lined with fire brick full height | 8 | 4½* | |
| Reinforced concrete or metal | | 4½* | |

\* Incinerators specially designed to produce low flue gas temperatures may be lined with fire clay flue liner.

Copyright © 1976, National Fire Protection Association — Reprocduced With Permission.

### Basement

Flooding in the basement can result from heavy rains or sewer backup. Ironically, the water is a fire hazard because it can cause electrical shortages in basement circuitry. Watch for pools of water or water pipe leakages in the cellar, and if they should occur, take immediate steps to find the cause of the trouble.

### Electrical Circuits

Protection from electrical overload should be provided by either circuit breakers or tamper-proof fuses.

Install enough electrical outlets. The standards of National Electrical Code state that any point along floor line is within 6 feet of an outlet.

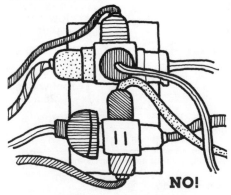

**NO!**

Outlets should be of the three-prong grounding type, or grounded so that grounding adapters may be used. A three-prong adapter may be used temporarily, but for frequent use, it's much safer to have a properly grounded three-wire outlet.

### Heating Equipment

Heating equipment should be constructed in compliance with safety standards and with proper controls to regulate heat output.

Heating surfaces or open flames should be located so as to avoid accidental contact with combustible material.

### Fireplaces

The hearth should extend at least sixteen inches into the room and eight inches on each side, and should have a damper.

### Surrounding Areas

Although we have been focusing on checking for the potential fire hazards within your home, danger can exist in areas nearby — in your yard, perhaps, or in surrounding land, or in the house across the street.

Sparks from an adjacent house could fly onto your roof and set your property afire. Your protection from this is to have a

roof and walls of fire-resistant materials as we have already discussed. In such a situation (besides lending a hand to your neighbor), close your home's doors and windows to stop flying cinders from igniting the interior. Contact the fire department yourself if you think no one else has sounded an alarm.

Your own yard can be improved by clearing it of any debris. The lawn and shrubs not only look better when they're trimmed, but they present less combustible material to a thoughtlessly tossed cigarette.

A real danger for those who live in the countryside is the possibility of brush and forest fires. In areas prone to these kinds of fires, you may want to cut a "fire zone" around your house by stripping away a circle of vegetation.

**A Special Note About Apartments**

Many of the problems and procedures that have been covered apply to apartment or condominium dwellers.

Because of the usual proximity of one apartment or condominium to another, however, the chance of a fire spreading to your place from a neighbor's home is far greater. In large apartment complexes with many living units, there will be a great many possible sources of exterior fire. And neighboring residents can generate fire hazards in areas of the building apart from the living space, for example by storing trash in common stairways or corridors.

The possibility of having less direct control over your living environment is no excuse for carelessness on your part. Apartment residents concerned about fire safety can still do a great deal to cut down on risk.

Keep a watchful eye out for fire hazards that may exist throughout a building. If you do find garbage strewn along public passages, report it to the superintendant or landlord. When you know of a resident in another apartment who creates an obvious fire risk — for example, by keeping gas burners turned on after leaving the apartment — politely ask the resident to stop the dangerous behavior. If the action persists, inform the landlord. Don't let a sense of "it's none of my business" intrude on your own safety.

Your lease should contain a guarantee that the building services are to be properly operated. A smoking incinerator, leaky plumbing, and faulty circuitry should be reported at once to the superintendant, who should be contractually bound to repair them.

The rules about having bedroom doors able to stand up to fire apply doubly to an apartment. The door to the outside hall should be tight-fitting and solidly made of fire-resistant material. At night, keep both the entranceway as well as bedroom doors shut.

You may wish to strengthen the walls and floor-ceilings separating your place from other apartments, expecially when residing in smaller apartment buildings. Studies of apartment fires have shown that the walls of low-rise buildings often cannot contain a fire.

Experts agree that smoke detectors are probably the best early warning system for fire that a home can have. Apartment dwellers might be wise to band together to ask the landlord to install smoke detectors in every hall. Putting up part of the money for installation would be well worth the expense. Of course every apartment should be equipped with at least one of its own smoke detectors.

**A Fire Department Check-Up**
No one in your community is likely to be as knowledgeable about fire prevention as the people who operate the hook and ladder. Inviting a few fire fighters over for a snack and an inspection of the premises could be very instructive. They could rate your house's fire safety, give advice on the escape plan you may have devised (see page 90), and tell you a little bit about how they approach the task of fighting a home fire.

Many fire departments have a regular practice of conducting home inspections. Ask the one nearest you what its policy is.

# Chapter III

# THE WELL-APPOINTED HOUSE
Home Fire Fighting Equipment

## Smoke Detectors — Great Protectors

A smoke detector, as the name would imply, is a type of warning device. This relatively simple, inexpensive piece of equipment has become very popular in the last few years because of its effectiveness in sounding an alarm before anyone sees or smells smoke.

It has been documented time and time again that the installation of smoke detectors can save lives. For example, in Fairfax County, Virginia, the fire prevention bureau recently reported that ten of twelve people who died in area fires might still be alive if they had installed smoke detectors in their homes. The three fires that resulted in the deaths all had "a distressingly common scenario."

In each case, incompletely extinguished smoking material ignited upholstery after the occupants had gone to bed.

A fire official commented:

"An operating smoke detector would very likely have awakened these people in time for them to escape."

Nine out of ten people killed in building fires die in the home. Often these fires start when smoking materials — ashes, lit cigarettes, or cigar butts — fall onto upholstered furniture or a mattress. The result is a slowly developing fire that may smolder for hours before bursting into flame. Smoke detectors sense the smoke and sound an alarm giving you time to escape. Most fatal home fires start between the hours of 11 p.m. and 6 a.m.

Fire damages property but the fact is that during a fire the flames are usually the last of the killers. Smoke and gases claim more lives than flames. Gases, hot air, and smoke can overcome sleepers before they can awake during the night.

The best defense is an automatic early warning smoke detector which can alert people to take action before the fire ever becomes more than just a small beginning-type fire. Smoke detectors dramatically increase the chances of waking up, getting out and calling the fire department before it's too late.

Many cities and counties across the United States require smoke detectors in new homes, apartments, condominiums. and mobile homes. Some communities have legislation requiring smoke detectors in both new and existing homes. The National Commission on Fire Prevention recommends that every home, apartment and mobile home have **at least** one of them, and tax and fire insurance deductions to householders who install them. (See page 108.)

The following list of tips on buying and using smoke detectors can help you get maximum protection from yours.

**Buying Smoke Detectors**
The smoke detector you purchase should be listed by a nationally recognized testing laboratory, such as Underwriters Laboratories. Its approval means that the detector is certified to meet basic performance standards.

There are two basic types of smoke detectors: photoelectric and ionization.

Photoelectric smoke detectors use either an incandescent light bulb or a light emitting diode (LED) to send forth a beam of light. When smoke enters the detectors, light from the beam is reflected from the smoke particle into a photocell, and an alarm is triggered.

The ionization chamber has a small radiation source that produces electrically charged air molecules called ions. These ions cause a small electric current to flow in the chamber. Smoke particles entering the chamber attach themselves to the ions, reducing the electrical flow. The change in current sets off an alarm.

**The presence of a radiation source in ionization detectors is not a hazard. The U.S. Nuclear Regulatory Commission checks into radiation saf ₃ty requirements before the detectors are placed on the market. Consumers Union has conducted independent studies which confirm that there is NO RADIATION DANGER from ionization detectors.**

Both types are approved by nationally recognized testing laboratories and either can do a good job in your home. Present technical evidence does not support a preference for one type or the other. The differences in response time are not considered critical for most residential situations.

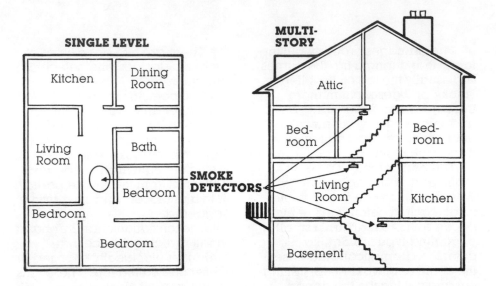

**SINGLE LEVEL**

Kitchen

Dining Room

Living Room

Bath

Bedroom

Bedroom

Bedroom

**SMOKE DETECTORS**

**MULTI-STORY**

Attic

Bedroom

Bedroom

Living Room

Kitchen

Basement

## Location of Smoke Detectors

A primary job of a smoke detector is to get sleeping people out of the house by warning them against danger. So, a smoke detector should be installed in the hallway outside each sleeping area in the home. In a house where the bedrooms are upstairs, the smoke detector should be near the top of the stairway to the bedroom area. The primary rule for locating a smoke detector might be stated as "between the bedrooms and the rest of the house, but closer the the bedrooms."

It is good practice to sleep with bedroom doors shut. A closed bedroom door can provide protection against fire and smoke from outside the room. The smoke detector in the hall should be close enough to the bedroom so people in the bedroom will be able to hear the alarm.

If there is a fire in the bedroom, however, the smoke will be prevented from reaching the detectors in the hall.

Therefore, it is a good idea to put a smoke detector in each bedroom, too.

Homes with basements should have a detector on the basement ceiling, above the bottom step of the stairway. About one-third of all fires in homes with basements start in the basement. It could be difficult to hear a basement detector from the bedroom when the door is closed. For this reason, it is suggested that homes with more than one smoke detector be equipped with models that can be interconnected. When one interconnected model senses smoke, all the detectors in the house sound an alarm.

Avoid positioning a smoke detector near the bathroom where water vapor from a shower or bath may seep into the hallway, because particles from the vapor can also activate the smoke detector at an inopportune time.

Since smoke rises, detectors should be placed on the ceiling or high on an inside wall, just below the ceiling. Avoid placing detectors in the "dead" air high in corners. You should not place smoke detectors within 3 feet of an air supply register or between furnace air returns and the bedrooms. For ceilings below uninsulated attics or in mobile homes, place detectors on an inside wall 6 to 12 inches below the ceiling.

In addition to the dead area referred to, avoid positioning near air vents which might prevent smoke from flowing into the smoke detector. If possible, try not to put the smoke detector on a ceiling where the air is of markedly different temperature from the rest of the room. Thermal barriers can prevent smoke from reaching the smoke detectors.

You may still have a question regarding the optimum placement of a smoke detector. You could contact the local fire deparment and ask one of their fire fighters for recommendations. Experienced fire personnel should be able to give you the right instructions. Otherwise get in touch with a representative of the detector's manufacturer.

**Installation and Maintenance**
Batteries or ordinary household current can power smoke detectors. Installation and maintenance vary for each type of power.

Battery-powered smoke detectors are easy to install; a screwdriver and a few minutes are all you need. Make sure that the detector you buy has a standard size battery which can

be purchased at most stores. **BATTERIES SHOULD BE REPLACED EACH YEAR.**

For the battery-operated smoke detector, it is a good idea to select a specific date when the batteries will be changed. Many smoke detector manufacturers suggest that a change every fourteen to eighteen months is adequate — once a year is a better recommendation. You might consider doing the change every New Year's Day; the important thing is to put "smoke detector battery day" on your calendar, and to take the time to perform the simple task of replacing the batteries.

Plug-in smoke detectors are equipped with an 8 to 9-foot electrical cord. Like battery-powered detectors, they are easy to install. Before purchasing a plug-in unit be sure that a wall outlet that cannot be turned off by a wall switch is available and can be reached by the cord from the detector.

Permanently wiring a smoke detector into the home's electrical system requires an electrician, who may charge between forty-five to fifty dollars for installation.

**How Reliable Are Electrically Powered Detectors?**
Electrical service is quite reliable in many areas, but less reliable in others. There is a possibility that your service may break down occasionally. **There is also a chance that an electrical fire will disable a smoke detector before an alarm sounds.** Since many electrical fires start in large

appliances, you can minimize this possibility by wiring the detector to a circuit that does not serve major appliances.

Some electrically powered smoke detectors have back-up batteries. Like the other battery-powered models, these units need new batteries yearly.

**Be Sure It's Working**
A smoke detector can't be a "great protector" unless it's activated. While the smoke detector should be tested at least once month, you should know that the unit's power goes down somewhat every time the unit is tested. A simple test for a smoke detector can be to hold a candle or lighter 6 to 12 inches below the unit. This should be more than enough to activate the system. Once the smoke detector has gone off, fanning the area around the smoke detector very gently with a newspaper or magazine will clear the chamber.

Some models have a test button that activates the detector. **All detectors should be tested monthly or when returning after the home has been unoccupied for a few days.** Read the manufacturer's instructions for complete details.

Some smoke detectors are equipped with a pilot light that glows or flashes to indicate that electrical power to the detector is on. Even detectors with this feature should be tested occasionally.

Approved battery-operated smoke detectors will sound a "chirping" alarm for seven days when the battery needs replacing.

Photoelectric smoke detectors give a trouble signal when the light bulb wears out and include a spare within the unit. Those photoelectric detectors using LED's (light emitting diodes) usually do not have a trouble signal for the light source due to the long life of LED's.

Most smoke detector manufacturers will back up their detectors with a service or replacement warranty. The design of most smoke detectors is relatively simple and trouble-free, but if a smoke detector does not perform during the warranty period be sure to take it back to the store where you made the purchase and have it repaired, or replaced entirely.

**A Great Gift Idea!**
One final suggestion about smoke detectors. Beause so many homes and so many families are not yet protected with this simple and effective device, you might consider giving it as a rather unusual yet practical gift. Smoke detectors make a fine present for holidays, birthdays, anniversaries, and weddings.

And what could make a more thoughtful housewarming gift than a smoke detector?

**Heat Detectors**
Heat detectors may act as effective complements to smoke detectors. Heat detectors are set off by the presence of nearby flames. They are best placed in closed areas where rapid, flaming fires are likely to develop. For example, heat detectors can be helpful in workshops,

basements, attics, and garages.

Your smoke and heat detectors are probably the most valuable items you can possess to protect yourself against the possibility of fire. Valuable as they are, however, they are only the **first** line of defense against fire. They alert you to the existence of trouble. Other pieces of fire fighting equipment can be considered preventive should a fire start in your house.

### Home Fire Extinguishers

Portable fire extinguishers are the most well-known fire fighting equipment that the homeowner, apartment dweller, or small business establishment should possess. They are light, easy to use, and very effective at putting out most kinds of fires. Fire extinguishers are designed to fight small, easily extinguishable blazes. It should be stressed that if you have any difficulty containing or putting out a fire with an extinguisher, you should leave the bulding right away.

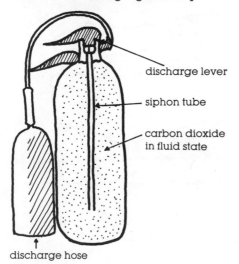

discharge lever

siphon tube

carbon dioxide in fluid state

↑
discharge hose

### Choosing an Extinguisher

There are several kinds of fire extinguishers which are designed for different kinds of fires. Fires are categorized as Class A, Class B, Class C, and combustible metal fires. Fire exinguishers have corresponding classifications.

Class A fires are those involving material like paper, cloth, wood, and many types of plastic and trash. The extinguisher recommended for this type of fire will have on it a capital A in a triangle. If it's colored, it will be green.

Class B fires result from the ignition of inflammable liquids, such as gasoline, kerosene, paint, or cleaning fluids. The extinguisher label has a capital B in a square and may be red.

Class C fires are related to electricity, and are usually caused by faulty wiring in equipment such as switchboards, wiring or an extension cord. The recommended extinguisher has on its label a capital C in a blue circle, if colored.

Combustible metal fires you need not worry about. They are rare, and occur when chips of certain highly flammable metals such as manganese are ignited.

In the average home, Class A and Class C fires, ordinary combustible and electrical fires, are most likely to occur. Class B fires, inflammable liquid fires, are less likely to happen, but are still common.

A water fire exinguisher is effective in fighting Class A fires. These extinguishers typically contain 2½-gallons of water stored under pressure.

Carbon dioxide or dry chemical extinguishers should be used against Class B and Class C fires, but they leave a residue that must be removed after the fire so the sprayed material won't be damaged.

It is best to keep several fire extinguishers, with one or more of each type, in different strategic locations around the house. If only one extinguisher is to be purchased, the dry chemical sort, rated A B C, is the best bet. An A B C extinguisher works with equal effectiveness on burning paper, flaming grease, and smoking television sets.

Extinguishers are also classified by the fire fighting strength. Choose the most powerful unit you and everyone else in your family can easily carry and use. Numbers from one to ten on the label designate the size of fire the unit can fight. If you decide to by an A B C extinguisher, get one having a minimum rating of 2-A: 10-B;C. (If you choose one unit for A-type fires, and another for B and C fires, the former should have at least a 2-A rating and the latter at least 5-B; C.)

**Operating An Extinguisher**
Learning to operate a fire extinguisher is a simple and easy task. The components of an extinguisher are few and uncomplicated. Take the time to study very carefully the directions on the label. Hold the extinguisher in your hands to get a feel for the weight. Get straight in your mind the few steps needed to activate the spray. Practice every once in a while to refresh your memory.

**← Side-To-Side →**

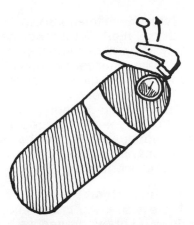

When using a water extinguisher against an ordinary combustible fire, direct the stream of water with a side-to-side motion at the **base** of the flames. Avoid fixing the spray at what appears to be the center of the flames. Continue spraying the affected area after a fire has been put out in order to prevent reignition.

**It's extremely important to understand that water should not be used on inflammable liquid or electrical fires.** A jet of water might spread the base of an oil fire, for example, or may further disrupt circuitry in an electrical fire.

You should use the same sweeping, lateral motion with dry chemical and carbon dioxide extinguishers. A locking pin is usually found on the handle of the dry chemical and carbon dioxide extinguishers. You can activate the extinguisher by pulling out the pin and squeezing gently on the operating handle.

Once a fire extinguisher has been used, take it immediately to the place of purchase to have it recharged. If the extinguisher is the kind that cannot be refilled, a new unit will have to be purchased.

**Locations for Fire Extinguishers**
There are specific areas around the house that should be considered in locating fire extinguishers. Look around for places where fires are most likely to start. The kitchen is an obvious danger zone. The garage, if inflammable liquids are stored there, is another; and possibly in and around a hallway closet and near sleeping areas. They should be mounted in an accessible spot near the room exit and away from such fire hazards as the kitchen range, or the workshop paint shelf. They should be placed out of the reach of children except those old enough to use them properly.

Some people like the idea of having a fire extinguisher in their automobile. A fire in the electrical system can be quickly extinguished before there's a chance of major damage to the motor, body, or, most importantly, the passengers.

Everyone in the house, including the smallest members of the family who will be expected to use it in an emergency, should be made aware of the location of each extinguisher.

Test fire extinguishers regularly to see if a charge has been maintained. Follow the directions on the label for the proper procedure.

## Sprinkler Systems

As more homeowners became concerned about preventive measures they could take to protect themselves from fire, the commercial sprinkler system was adapted for residential dwellings. Similar in design to the models used in factories and office buildings, the home sprinkler system provides precise fire suppression.

The sprinkler system is simply a separate series of pipes with strategically placed sprinkler heads. A single properly located sprinkler head often can spray enough water to put out a fire in a fair-sized room.

The system is controlled by a heat-sensitive "fusible length" attached to the heads. When the fusible length detects abnormally high temperatures in a room, it activates the sprinkler. While temperature ranges vary among fusible lengths, a temperature of 130° F. will usually result in a room being covered with a spray of water. Since 130° F. is well below the temperatures produced by a fire, sprinklers will serve as a very early warning of ignition.

The ideal time to install sprinklers is when a building is constructed. The cost of putting in a separate plumbing system at a time when the main plumbing is installed is fairly low.

Building a home sprinkler system in existing structures, and particularly in older homes, can be costly. The expense may be less if an insurance company gives breaks on the price of its fire insurance or homeowner's policy in exchange for installation of sprinklers. It is prudent to ask both the installer and insurance firm about their rates before making a decision.

## Baking Soda Pail

Recently the folks at *Arm and Hammer* came up with an interesting variation on an old firefighting tool. They suggest filling up an emergency fire pail with baking soda.

Small, fast-starting grease or electrical fires take place with alarming frequency, especially in and around the kitchen. Packing a pail or can with baking soda gives you a light, refillable, and easily manageable weapon against these flash fires.

Baking soda or sodium bicarbonate is a cheap food product found in any grocery. It smothers fire because it releases carbon dioxide when heated. Most people who have used it on fires poured it out of a box. This can be awkward, as a person has to get dangerously close to, or even over, the flames to drop the baking soda. Also, the spout on the box is rather small.

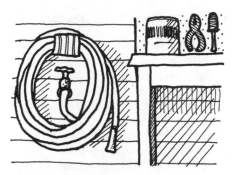

You can easily get around this little problem by keeping the baking soda in a 1-pound coffee can. Attach a large handle to the can and you're ready. You can even glue a 'fire pail' label to the side to make it look official. The plastic lid that comes with the coffee can can be flipped off in case of a fire, and will keep the baking soda fresh and dry.

This handy device can go in other places besides the kitchen. Put it in the garage and laundry room too, and set it beside the grill during barbecues. You can use other substances instead of baking soda. Regular old sand will fill a can nicely, and smother flames too. Remember, however, that your fire pail will only suffice for small fires. It is not intended to take the place of fire extinguishers.

## Hoses
One means of fighting fires that many homeowners already have is the garden hose. A hose hooked into the conventional water supply can be very handy in dousing exterior fires, garage blazes, or in wetting down shrubbery threatened by a fire outside the immediate property. Unlike fire extinguishers, hoses will spray indefinitely. If you already own a hose, look into getting an extension. The extra length will increase the range of this often overlooked but handy tool.

## Escape Ladders
Sometimes a fire cannot be fought. When a conflagration rages out of control the only proper response is flight. For the family living in a single-story ranch house or on the bottom floor of an apartment, access to the outside (provided escape routes aren't blocked) should be fairly easy. It is a different matter for people living on the second, third, or even the fourth floor of a dwelling. If the only way out is through a window, they might have to face a considerable leap to the ground.

An escape ladder is a solution to the potential dilemma. The escape ladder is normally positioned inside one of the exit windows. In the event of an emergency the ladder can be slid out the window and down the side of the building. Alternately a ladder may be fixed permanently on an outside wall.

Practice drills are essential if escape ladders are to be used safely. Even for a young and agile person, climbing down the side of a darkened building at a time of distress can be a trying experience. It might be well-nigh impossible for an elderly person to accomplish such a feat unless thoroughly familiar with the right procedure.

A portable ladder should be kept next to the exit window. A rule of thumb is one escape ladder for each sleeping area with an available exit. The portable kind can be broken down and stored in a small metal cabinet. Refrain from using the cabinet as a stand for household items like plants, linen, and books. Keep it clear and ready for emergencies. When drilling house members in the use of the ladder, make sure they know how to extract it from the cabinet and expand it to full length.

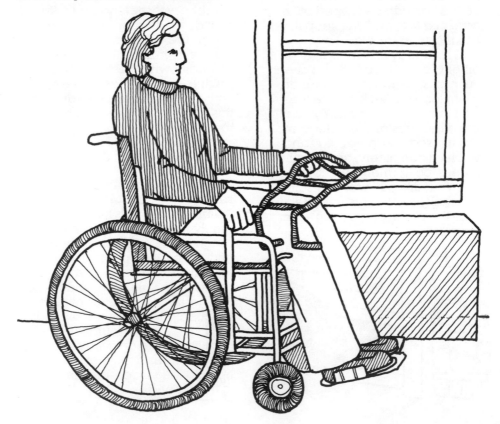

### Flashlights

Life-threatening fires most frequently break out at night, and may disrupt the electrical circuitry that powers the house lights. To move about in the dark you need a light source. Open flames like candles and matches carry obvious hazards. Therefore flashlights are essential for the well-appointed house. Flashlights should be placed at key locations all around the house. At the absolute minimum they must be kept in the bedroom. A periodic check of the batteries is mandatory.

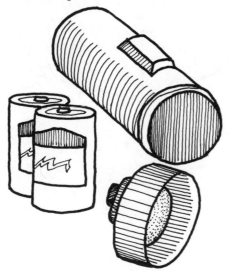

### Automatic Telephone Dialer Systems

The telephone company in some communities will offer a service in which a smoke or heat detector is hooked up to a telephone dialing device. If a detector is activated the phone system automatically dials the fire department. A tape recorded message giving the house whereabouts is played.

On numerous occasions fire fighters alerted by an automatic dialer have arrived at a house to handle a fire **before** the residents of a house had become aware of trouble. Residents had not heard the siren of the smoke detector hooked up to the phone. Automatic dialers are a bonus in a crisis situation where there is no time to place an outside call or where the phone number of the fire deparment is misplaced or forgotten.

Check if an automatic telephone dialing service is available in your area.

### Exterior Horn

A horn or siren attached to an outside wall can be hooked up to a smoke detector inside. If the detector goes off the horn will too, alerting everyone in the vicinity to trouble. A horn is added protection, for example, for the family that travels a lot. Of course, a horn's effectiveness hangs on how cooperative and attentive one's neighbors are. If they are an extremely apathetic sort (or if, for that matter, you live in a very isolated setting), a horn may make a poor investment.

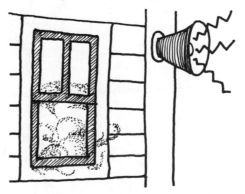

## Extra Clothes

Residents making a swift escape from a burning house may have had no time to dress properly before leaving. This could pose problems if a fire happens on a cold winter night. Especially if your dwelling is located in northern climes, it may be wise to store some old clothes in a separate building like a garage. Storage of footgear would be most appropriate. Imagine having to stand out in a half foot of snow without any shoes while you wait for the fire engines to come.

With smoke detectors, heat detectors, fire extinguishers, and sprinklers installed and in working order. With baking soda handy, hoses at the ready, escape ladders, flashlights that light, a telephone dialer system, and extra clothes stashed away, you have a very well-appointed house. If you can't have it all, choose smoke detectors — first things first.

# Chapter IV

# **Everywhere You Roam**

## Fire Safety Away From Home

### Be Vigilant and Prepared

Up until now this book has dealt with fire prevention in conventional homes and apartments. But the incidence of fire is not restricted to family frame houses or garden apartments. Flames can strike any living quarters, as well as places of work and recreation. Anywhere you go, you should know how to obey the rules of fire protection.

### Hotel/Motels

Hotel and motel fires have always gathered plenty of headlines. Recently there has been a spate of sensational news stories about major, tragic fires in large, nationally known hotels. The publicity given these blazes has had the good effect of making fire safety a genuine concern in the minds of travelers and hotel proprietors. For example, it is not unusual today for management to install smoke detectors throughout a hotel. It is not unheard of for vacationing families and people traveling on business to ask for rooms on the lower floors to permit a swifter departure in the event of fire. The Federal Emergency Management Agency of the U. S. Fire Administration recommends that you ask if the hotel or motel is equipped with a sprinkler system or at least has smoke detectors in the rooms when planning your trip. It's a good idea to pack a flashlight and a smoke detector in your suitcase when you travel. The flashlight may be useful to light your way in an emergency. The placement of the smoke detector in the hotel room, should be in accordance with manufacturers' recommendations.

It has been shown that a great many hotel and motel fire deaths are caused by toxic smoke inhalation.

Surviving a hotel fire can be no problem if upon entering a hotel you take a few minutes to check a few basic items.

First, find out what the fire alarm system is like, and ask what it sounds like. One big hotel had an alarm that sounded like Big Ben at Westminster and no one paid attention when the chimes began ringing.

### When You Check In

Take a walk down the hall from your room and count the number of doors in both directions to the nearest fire exit staircases. Check the door to the exit stairs to be sure it will open if necessary, but closed at the moment.

Glance at the stairs to see if they are free of rubbish or other obstacles.

Locate the nearest alarm box, and read the instructions so you know how to set it off quickly.

Next, look about your room. Study its layout. Many people have lived through a hotel fire when escape routes were blocked off by staying in their rooms, where they were protected from smoke and gases which choked off a corridor but failed to penetrate a closed door.

Try the windows to be sure they can be opened. Decide which one you would use in an emergency.

Look out your windows to see what is outside. If you're on the upper floors, there may be an escape ladder, or a deck within safe leaping distance. If the hall is not usable, you may want to jump, but dropping more than two floors probably would result in injury.

Put your room key on the bedside table while in the room. If there is an emergency and you find the exits blocked, you'll need the key to get back into the room.

Report any fire safety deficiencies to the front desk and read any posted fire emergency information.

## FIRE IN YOUR ROOM

Leave the room, and close the door to keep the fire from spreading. Alert others by pulling the nearest fire alarm.

Proceed to the nearest stairway exit. Remember, DO NOT USE ELEVATORS. Notify the front desk.

Don't try to extinguish the fire unless you know you can contain it.

## FIRE IN THE BUILDING

Follow the posted emergency instructions.

To determine if it is safe to leave the room, take your room key, and go to the door. If there is any smoke in the room, and you're in bed, roll out onto the floor and crawl to the door. Don't stand up. Feel the room door with your hand. If the door feels cool open it a crack. If there is no smoke, walk quickly to the nearest exit after closing your room's door. If there is smoke, crawl down the hallway counting the doors to the nearest stairway exit. The air should be most breatheable about 12 inches above the floor. If this exit cannot be reached, try getting to the other fire exit. If getting to the other fire exit is not possible, turn around and count the doors to get back to the room. Unlock the door and enter.

If you are able to reach the exit, walk down to the ground level. DO NOT USE THE ELEVATOR. If fire or smoke is dense at lower levels, go up to the roof, prop open the door to vent the stairwell and protect yourself from being locked out while you wait for help.

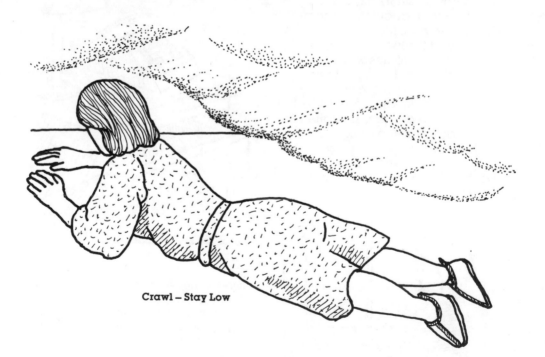

**Crawl – Stay Low**

## TRAPPED IN THE ROOM

If it is unsafe to leave the room during a fire, several precautions should be taken.

Turn on the bathroom fan if there is one. The fan will vent smoke entering the room.

Fill the bathtub with cold water. Use this water to wet towels or wet the door down. Stuff wet

towels or clothing under and in the sides of the hotel or motel room door to keep the smoke out. Your ice bucket can be used to bail water if necessary.

Close as many doors leading to the fire as possible. Try to remain in the room that has a window.

If there is no smoke and gases rising up from a window on a lower floor, open the window to let out the smoke.

Let somebody know where you are. Use the telephone if it is in operation. Hang a white sheet out the window to signal fire fighters but don't try to climb to the ground on a string of tied together sheets.

If the room gets smoky, stay low. A wet towel should be held to the face. DO NOT JUMP. The fire department is on the way.

If smoke makes breathing difficult, lean out a window with a blanket acting as a tent draped over your head to get fresh air. You may have to break a window with a chair or drawer.

If it becomes impossible to stay in your room and you are forced to make for the exit, remember to keep low.

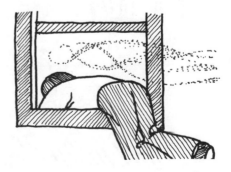

## Workplace

Extensive research has been done on fire in the workplace. The main causes of work-related fire are careless smoking, electrical hazards, and the accumulation and failure to dispose of combustible materials.

Smoking has resulted in more occupational fires than any other single factor. Employees smoking cigarettes have often touched off explosions in work areas containing flammable gases or liquids. "No smoking" signs which are easy to read should be positioned around places where smoking is dangerous. Many fires are started when a worker carelessly tosses aside a match or cigarette. Employees should be provided with plenty of ashtrays and be encouraged to use them — even better, to stop smoking.

Electrical hazards are the second greatest cause of business fires. A semi-annual maintenance check of wiring in the office or factory is mandatory. Workers operating electrical equipment should repair tools exhibiting signs of wear. Unnecessary lights should be turned out at the end of a shift.

Combustible materials must be sealed in the proper containers. Waste matter should never be left exposed to the air. Containers should be regularly inspected for leakage or signs of wear. Combustible materials should be stored in an area accessible to authorized workers.

Sizeable businesses sometimes contain large kitchens and lunch rooms. Facilities ought to be kept neat and orderly. Fire extinguishers should be positioned

around cooking apparatus, which must be inspected for gas leaks and efficient ventilation.

In drawing up a fire protection plan for a place of business, it may be helpful to call on expert outside advice. Invite members of the local fire department to visit the establishment several times a year. Fire officials can perform a professional estimation of the workplace to identify any violations of fire and safety codes. They are in the best position to make suggestions on ways of eliminating fire hazards.

The fire officials invited over to a business know the ins and outs of fire dangers in the workplace. But eliminating hazards is only half of good fire protection. Fire officials can also offer counsel on the best way a business can react to a fire in progress.

For example, a fire department can help a business set up a fire escape plan. This would entail finding out the routes by which employees could make their way safely out of a burning structure. Copies of escape route diagrams would be distributed to every employee and posted on bulletin boards. Main fire exits would be plainly marked and major escape corridors kept free of obstacles.

Fire drills are an essential part of any organization's fire reaction plan. There is no reason for elementary and high schools to hold fire drills regularly while institutions and companies ignore the practice. Drills are a check on whether employees know their escape routes and on whether fire alarms are functioning properly. Fire drills should be held at odd intervals throughout the year.

A large fire can ruin inventories, wreck expensive equipment, and put employees out of work. A smart business manager keeps firefighting equipment on its property to prevent huge fire losses.

A sprinkler system is required by law in some establishments. Sprinklers may well be worth paying the price of a significant installation fee because they provide an effective and almost instantaneous response to the outbreak of many fires.

Sprinkler systems may be required by law but maintenance schedules may not be. Sprinkler heads must be kept clean and free of corrosion. The heat-sensitive device on a sprinkler must be operating. An adequate water supply must be assured, as well as conduits for getting water to the sprinklers. Water valves must remain open: fires have raged unchecked in buildings with the highest quality sprinkler systems because the valves controlling water flow were closed.

Fire extinguishers should be distributed throughout the workplace, within easy reach of any employee. Extinguishers at work or home must be inspected regularly to be sure they are in good working order. Workers should know how to put out a fire with an extinguisher and should practice the technique every few months.

If there appear to be dangerous conditions at the place of work these should be reported to the Occupational Safety and Health Administration.

Finally, the best early alert for fire is a smoke detector. Smoke detectors placed in major corridors and in areas susceptible to fire will in many incidences allow employees to clear out of a building before their lives are threatened.

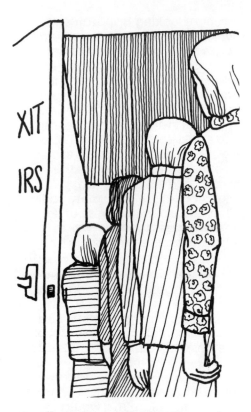

### High-rise Buildings

Many of the principles previously touched upon regarding the best way of responding to fire in hotels, apply to high-rise build-ings of all kinds. The temptation to use an elevator must be resisted if there is a fire. Since you do not have complete control over an elevator's destination, the elevator may bring you to a floor where a fire is burning out of control. Fire-safety experts agree that you should use the stairway even if it is a long hike down.

Particularly in the case of large buildings, a place to meet outside in the event of a fire must be agreed upon in advance. A

meeting place is the best way of determining whether everyone concerned has gotten out of a burning building.

One of the major fire problems with high-rise office buidings today springs from the use of central heating and air condition-ing. Since temperatures in modern buildings are completely regulated, some designers thought they could do away with windows that open and close. Permanently sealed windows were installed purely for appearance. Of course, windows that cannot be opened during fires can turn rooms into fire traps. Don't be hesitant about breaking a window, sealed or otherwise, if smoke starts to fill up a room.

A clear advantage most high-rise buildings have over their smaller cousins is greater fire-resistant properties in the walls. Walls, ceilings, and floors in big buildings usually stand up to fire for longer periods. Thus, in tall structures you may have a greater chance of riding out a fire by staying in your office or apartment.

## Dangers of Rubbish

Office and factory workers and apartment tenants seldom think about cleaning up the space around their building. Yet accumulations of rubbish and weeds outside an office or factory can constitute a real fire hazard. Maintenance people should clear away accumulated debris and continue the upkeep of building surroundings. Follow the safe practices listed below:

— Clear rubbish away from loading docks and window walls. Rubbish stored temporarily in containers should be placed well away from the building.

— Have rubbish removed regularly. This may involve contracting a private disposal firm to supplement the work of the municipal sanitation department.

— Do not burn rubbish in the open or close to the building. Incinerator stacks must be positioned away from main building sites.

— Store combustible items (cartons, pallets, papers) at least 25 feet from a building.

## Rural Houses/Vacation Houses

Residents of rural dwellings must pay particular care to dealing with an outbreak of fire. Houses in thinly populated areas may not have the benefit of a nearby fire department. Therefore residents may have to rely on themselves to extinguish a fire.

If there are no adequate sources of water available a water storage tank can be constructed. Fire hoses and a pump able to direct large amounts of water against a blaze should be purchased.

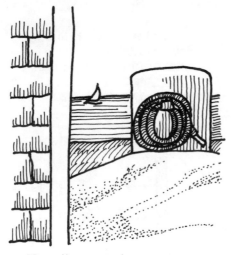

water down surrounding brush in very dry years. A wide swath of plain earth should be cut around living places. Clear away vegetation growing on the sides of buildings. Vines are picturesque but combustible, and thus constitute a fire hazard. If you're keeping piles of logs for the fireplace, keep them covered and apart from structures.

Vacationers have an unfortunate habit of taking the condition of summer homes for granted. In actuality, places occupied for just a portion of the year fall off in maintenance. So, in moving into a temporary home, take care to see if everything is in order. Clear debris away from doors and cubbyholes. Determine if windows can be easily opened and shut. Check in particular the utilities — is the furnace working okay, does the stove give you trouble, can you get plenty of water from faucets?

Wood and coal stoves are making a comeback in conventional houses. It might be said they never left rustic dwellings. These stoves are quaint and usually do a sufficient job, but are a greater fire threat than modern conveniences.

Vacations are for, among other things, relaxation. However, time off from work is not a license to let down your fire guard. If you travel to a country house to spend a week or two, don't neglect to bring along a smoke detector and fire extinguisher. Be as careful about fire as you would be at home.

Rural houses are more likely to be planted in the midst of thick vegetation. You may want to

Covers should be tightly fitting, flues should be clean, and fuel must never be placed near the stove. Don't use a highly inflammable liquid to get a fire going.

## Mobile Homes

The mobile home can be an exciting way to see the country. For many people a mobile home or trailer furnishes an inexpensive opportunity to own a house. Yet enthusiasts should know that the relatively small interiors of such homes allow fire to spread at an extremely rapid pace. Therefore smoke detectors are a reliable safeguard even in small recreational vehicles; the placement of one or two detectors in a mobile home is even more essential. A fire extinguisher is standard equipment, too. But remember how fast fire will expand in a closed environment. If your first efforts at quenching a blaze are unsuccessful, further efforts will likely be futile. Leave the trailer as fast as you can.

You can avoid a lot of potential trouble by buying a good trailer. The cabin should have multiple entrance and exit-ways. Wall and floor coverings should be made of material which a fire cannot quickly consume. The vents of large appliances should lead to the outside. Lastly, maintain the motor very well. You potentially have a lot more to lose from a fire in a mobile home's motor than in an engine of an auto.

## Houseboats

Large boats share with mobile homes the dangerous characteristic of having narrow, enclosed interiors. And with houseboats it is hard to find professional assistance to help fight a fire. (Fire tugs are extremely rare on lakes and the high seas.) So approach a big vessel in the same careful way you approach a mobile home.

Boat passageways are hard enough to move through without being cluttered as well. Keep them free of debris. Engines should be well maintained and separate from the other compartments.

The cabin that holds the cooking equipment should be thoroughly ventilated. All equipment should be operating properly, in "shipshape" condition. The stove itself must be of the kind approved for marine uses. Never prepare food on board ship in bad weather.

Steer away from stormy seas. Apart from the potential of capsizing a boat, a storm may carry the danger of lightning.

Everyone can learn to use prudence outside the home as well as in it. Daily living habits such as locating the fire exits when entering a building or a restaurant can, in time, become ingrained. It is the right — even the duty — of all of us to protect ourselves and others from fire.

Chapter V

# It Pays to Prepare

In Case of Fire

**Sometimes the most pains-
taking and thorough fire prevention isn't enough to stop
the outbreak of a fire.** Accidents happen even in the most
carefully managed homes.
Certain dangers may get overlooked, and even equipment
that has been recently checked
can malfunction.

In addition, posession of the
most high-quality fire detection
and fire fighting equipment may
not suffice in halting the spread of
flames. Despite the best efforts of
residents, a fire may start and
rage out of control. The residents
may be forced to flee the house.
In such an eventuality they will
need a plan of escape.

The same care, thought, and
foresight that goes into
eliminating fire hazards from a
home should be spent on
drawing up an escape plan.

An escape plan is vital
because in a fire every second
counts. This is especially true at
night, the time when most fires
occur. Families are most
vulnerable after the sun sets.
Flames may spread throughout
the house before anyone awakes
to the danger. People roused
suddenly from sleep by fire are
likely to be groggy, less alert than
normal. A fire may cut off the
electricity, which might make it
extremely difficult to move about
safely. A well thought out plan
will take much of the confusion
and uncertainty out of such a
situation, and allow a family to
make its escape in the fastest
possible time.

The National Fire Protection
Association has provided Operation E.D.I.T.H. (Exit Drills In The
Home) as a family plan for home
fire safety. The program outlines
three basic steps of fire readiness:
(1) have an escape plan, (2) buy
and install smoke detectors, (3)
practice your plan.

## HOW TO MAKE THE ESCAPE PLAN

### Draw a Floor Plan

The first step in putting together
an escape plan for a house is to
obtain a floor plan of the house or
apartment. A good floor plan
shows the windows, doors, halls,
stairways, and rooftop that could
provide means of escape. Every
household member should have
a copy of the floor plan and
should memorize it.

Locate all doors and windows
that can be used as exits. The best
and primary exit is the one that is
used most of the time — the front
or back door.

### Learn Escape Routes

Fires can spread through a home
with amazing rapidity. Flames
and gas may shut down a planned exit, so it is important to plan
at least two escape routes. One
route would be the way in which
a person leaves the building
under normal circumstances. The
second route is the alternate,
emergency route. It might involve
climbing out a window or going
down the back stairs.

On the next page is a typical
apartment floor plan with escape
routes marked

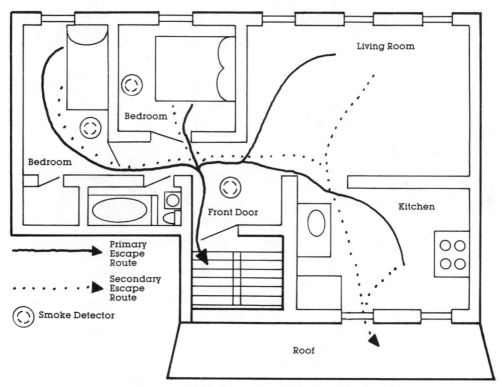

**An Escape Plan For A Typical Apartment**

**Establish a Meeting Place**

Another key part of an effective escape plan is to have a single, pre-designated area where the entire family or household will meet after they have escaped from the house. In addition to making absolutely sure that everyone is safely evacuated from a burning structure, this strategem also prevents any unnecessary and possibly extremely dangerous attempts by fire fighters to reenter the house. A meeting place will ensure that fire personnel know who is and who is not accounted for.

**Know How To Report The Fire**

It is a matter of civic duty as well as personal protection to notify the local fire department about a fire as soon as possible. After all, unchecked flames could spread from your house to others in the neighborhood. Fire information must be transmitted promptly and accurately.

Chapter III contains information about an automatic telephone dialing system and an exterior horn. If you alert the authorities yourself from inside your home, obey the following instructions:

— Make completely sure that all emergency telephone numbers (fire, police, hospital) are next to every phone.

— At the earliest possible age, teach your children how to dial the emergency numbers or instruct them in how to dial the operator and ask for assistance.

— When reporting an emergency, remember to speak slowly and understandably under the most difficult conditions. Stay calm as your call for help will do little good if no one can understand you. Give your name and address. If the house is difficult to locate, it might be helpful to give brief directions.

— Tell the operator or fire department dispatcher the kind of assistance you need.

Since a fire can disrupt telephone lines or cut off access to a telephone, it may be impossible to call for assistance from inside the home. In this event call the authorities as soon as you make good your escape. Know where the nearest phone or alarm is outside your home. Tell your children how to set off an alarm box. You might plan to use a neighbor's phone as a back-up.

### Have a Discussion About the Plan

Once the escape plan has been drawn up, and the diagrams with the alternate emergency routes are completed, get the household together to discuss the escape plan. Everyone must leave the meeting knowing at least two ways out of each bedroom, and also where the meeting place outside is located.

Emphasize that everyone's exit from the dwelling should be calm and rapid. No one should waste time looking for a cherished toy, pet, or article to take along.

When everyone has learned all the details of the escape plan, it is time to schedule the most important procedure of all, namely the escape drill.

### Practice the Plan

Make the fire drill as realistic as possible. Have everyone get into bed, lights out, with doors safely shut, and then set off the smoke detector. Everyone should then leave the house by way of the primary escape route.

A realistic drill will shake the "bugs" out of escape plans by making individuals realize the mistakes they made..Practices will make residents learn to keep a flashlight, for example, or a pair of shoes next to the bed to better prepare themselves for the next time the alarm goes off.

As they make their way outside, residents should remember to stop by a phone, if one is on their route, and feign an emergency call. Emergency personnel must be informed promptly of a real fire.

Everyone should meet outside at the prearranged spot to count heads. Always pretend that no one was able to get to a house phone in order to review the stand-by plan for getting in touch with the authorities.

Try the secondary escape route at the next drill. Practicing the alternative is important because the escape may be by way of a hard-to-traverse route such as down a ladder or across the side of a roof. If anyone has difficulty opening windows during this drill, take notice and solve whatever the problem is.

Time the escape with a stopwatch the first time through, and anytime thereafter. Cut down on the time needed with each session. Be the fastest, and safest, household on your block.

Here's a suggestion: Involve your children in this program by allowing each family member an opportunity to call a surprise fire drill within a week or two after you have made your plan. Be prepared, however, for Junior may call his first drill when Sis is in the bathtub.

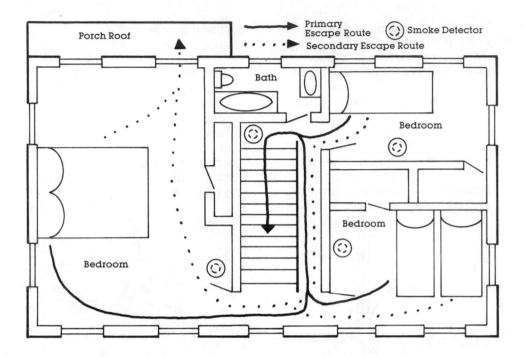

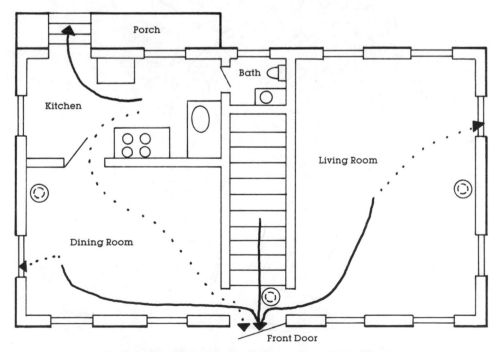

**An Escape Plan For A Typical Two-Story House**

## WHAT TO DO IF FIRE STARTS IN YOUR HOME

### Early Warning

The escape plan assumes, even demands, that you'll be alerted soon after a fire begins by an early warning device. Don't assume that you'll wake up instantly to the smell of gas or smoke (some gases given off by combustion have the effect of inducing sleep), no matter how light a sleeper you may be. A device like a smoke detector is essential to add true margin of safety to your house.

### Act Fast

When fire strikes, it's extremely important to act fast and efficiently, primarily because you may have only a couple of minutes to safely escape. This element of speed should be drilled into house members, and continually improved upon, during practice escapes. Don't waste valuable time trying to locate the fire. You may have to forsake calling the fire department from inside your home.

Your first responsibility is to get everyone outside to the predetermined location and then, and only then, should you send someone to a neighbor's house to call the fire department. Insist that everyone stay at the agreed upon spot. Don't let anyone wander off. Under no circumstances should anyone be allowed to go back inside the building to retrieve a posession.

### Attempt The Primary Escape Route

If the path is clear make way by means of the primary route of escape. It would be wise to wrap a towel or cloth around your face, in case you encounter light smoke in the hallway. You should never attempt to walk or crawl through dense smoke.

Often the first step in implementing the primary escape route is leaving a bedroom. The utmost care should be exercised in opening a bedroom door. If the door is tight-fitting — as it should be — you may be unaware of smoke and flames on the other side.

The first thing to do is to gingerly feel the door. If it is hot, don't open it! If you do, super-heated gas and smoke will likely rush in and strike you down instantly. Instead of opening a hot door, leave through a secondary exit. If the room lacks a window or another secondary exit, you will have to wait there to be rescued. If smoke is filtering through, stuff the cracks with clothing. Then get down close to the floor (where the air will tend to be fresher) and await the arrival of help.

If the door feels okay to the touch, open it very cautiously. Be positioned in such a way as to quickly slam the door shut if flames are discovered.

If the corridor seems safe, proceed outside. Rouse the other house members if you can. And leave as quickly as you're able.

## Attempt the Secondary Escape Route

The typical exit for the secondary escape route is a window. Here you can run into all sorts of unexpected problems.

A window has to be big enough, first of all, for a person to fit through. Alternately, in a child's room the sill has to be low enough for a child to reach. A step ladder can be kept by the window if this is the case, or plans could be made for the child to move a box or a light trunk to the window to use as a steppingstone to safety.

Some people have an alarming practice of nailing or painting shut windows. Don't do this. Windows should be easily opened in the event of an emergency. Practice escapes will uncover windows that are hard to open.

In a critical situation it may be unrealistic to take the time to pry open a storm window or a screen. Residents should know how to smash a heavy object against such barriers in order to force an opening.

If a bedroom does not have a window, an effort should be made to link it to a room that does have a secondary exit. Cutting a connecting passage through an adjoining wall is not an extreme action to take if it results in opening up another alternate escape route.

So far in this discussion the assumption has been that a window is not far off the ground. Obviously a second or third story window may be too high from ground level to leap from.

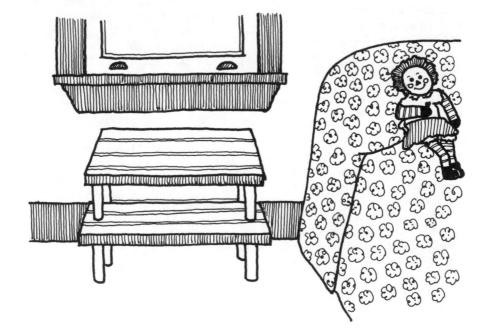

If an upper floor window leads out to a roof ledge, there may be no choice in a fire but to go out onto the ledge and stay there until rescue squads come to the scene. If a window drops straight off to the ground, however, without any intervening ledge, you should consider getting a rope ladder.

A collapsible ladder can be kept next to a window for emergency use. Unfolding or putting together a ladder can be a tricky business, so have anyone who might need to use a folding ladder practice with it. The ladder must be of the type that can be quickly and securely attached to the window.

It takes time to learn how to handle a collapsible ladder, and more importantly it takes time to get one ready during a fire. Therefore you may want to

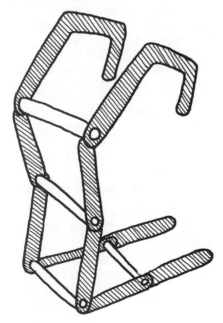

permanently attach a ladder to the appropriate outer walls of the home. People can also scale down the side of the building by means of a thin metal pipe. You must be rather agile to accomplish the latter feat. Special arrangements should be made for aged or disabled persons.

**What To Do If Trapped**
Leaving a house quickly is the ideal response to a fire. Yet under certain conditions you may be left with no other choice but to stay put in the house and await the coming of the fire department. A fire, after all, could block the primary escape route, and a secondary escape route may not exist. Even with unfortunate circumstances such as these, you can vastly improve your margin of safety — your ability to withstand fire until you're rescued — by following a few rules about fire behavior.

The first thing is to know your enemy. In waiting out a fire your major adversary is likely to be smoke. The key to surviving smoke is the fact that smoke tends to break up into layers.

Let's assume a hypothetical situation. Say you are trapped in a bedroom four stories off the ground. The room has one small window, but no means of climbing down to the ground. The only escape route is through the door, but in attempting to open the door you found there were flames in the outside hall. You slam the door shut, and smoke starts to drift in through the cracks. What should you do?

Remember that smoke does not equally distribute itself throughout an enclosed space. It will probably stratify into two layers in your room.One layer will hang near the ceiling, and the other layer will be close to the floor.

You should find relatively smokeless air about 12 to 18 inches off the ground. Crouch low to take advantage of this air. If there is a towel or article of clothing available, press it up to your mouth and nose. The material will screen smoke particles. You may have to crawl away from a part of the room which has denser smoke. Get on your hands and knees and move quickly away.

That little window can be of assistance now. Try opening the top and bottom of the window a little bit. Some of the smoke will pass out the top. Outside air will come in through the bottom.

Try to remain as calm as you can. Hold your breath occasionally to reduce the possibility of smoke inhalation.

## What To Do If Clothing Catches On Fire

One of the most difficult subjects to address in dealing with home fire safety is what should be done if a household member's garments catch on fire.

The one thing a person must not do if his clothes ignite is run about. The reason why running is dangerous is that fast motion will draw more oxygen into contact with the fire. The fire will burn more intensely, spread over clothes more rapidly, and cause more severe burns the more oxygen it consumes. Don't panic. Handling the situation calmly will result in less severe injuries.

For years specialists in the field of burn treatment have been telling the public that when clothing catches fire, the best procedure is to stop moving, drop to the ground, and roll over repeatedly until the flames go out. If the clothes of another person are flaming, you can use a blanket to smother the flames. After a clothing fire has been put out, fabric may continue to smolder. Smoldering clothes should be taken off immediately. However, if burned fabric adheres to the skin, don't try to take the fabric off. Follow the instructions for treating burns in Chapter VII.

In discussing the house escape plan, everyone should be instructed in the right technique of handling burning clothes. It would be constructive if community organizations such as schools, churches, and police and fire departments held public demonstrations of the technique. Children as young as five and six years old can be taught an invaluable lesson — STOP, DROP AND ROLL!

Remember these three words:

1. **STOP!** Don't run, Running fans the flames and causes them to burn faster.

2. **DROP!** Cover your face with your hands, then drop to the ground and keep your head out of the fire. This protects your

**STOP!**

**DROP!**

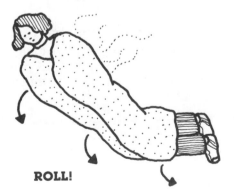

**ROLL!**

respiratory system, face and eyes.

3. **ROLL!** Grab anything available, such as a blanket, coat or rug, to smother flames and keep rolling in it. If nothing is available, keep rolling on a surface such as dirt, grass, vegetation, etc.

### Leave It To The Professionals

Assume that an escape attempt has been successful. All house members are safely accounted for outside the building. Fire engines and police cars are en route. A blaze is raging in the building.

You may get the idea to attempt to extinguish or limit the blaze with the help of a garden hose or home fire extinguisher. You may reason that it is a citizen's duty to give safety personnel a helping hand.

The fact of the matter is that amateur fire fighting is unlikely to be much help in stopping a blaze. In fact, many times it has happened that an amateur with good intentions succeeded, through ignorance of correct fire fighting methods, in spreading a blaze. Of course, many people have lost their lives in the process.

Trained fire fighters most probably will not need your help. Fire fighters are professionals who know how to limit the spread of a fire, and protect a home's interior from undue damage.

## If You Live In An Apartment Building

Escaping an apartment fire should be a similar procedure to leaving a hotel or motel. There is an important difference — and advantage — familiarity with your home turf. You should know where there are two escape exits other than elevators. (In a fire elevators may not be accessible, or may stop if power fails. They could trap you.) You should know, too, how to sound the fire alarm and call the fire department. Your family fire escape plan should be posted where everyone can see it. This plan should be rehearsed so that every member of the family could find the exits even in darkness, and would react automatically in case of fire.

This chapter emphasized that every good escape plan includes a place where every resident of a household can assemble outside. A meeting place is vital with an apartment fire, because the sheer number of residents, observers, and firefighting personnel milling around outside the structure will be much greater. It will be impossible to find people in the confusion unless everyone in a household goes to the same rendezvous point.

If there is a fire, here's what you should do:

1. Get your family out of the building. If caught in smoke, keep low where the air is better. If possible, cover your face with a cloth. Take short breaths through your nose.

2. Call the fire department. Only if the fire is very small, and you are able, should you attempt to fight it.

3. Feel the doors before opening them. If they are hot, use the alternate fire exit, or take refuge in a room with an outside window. Close as many doors between that room and the fire as possible. Open the window at the top and bottom and hang a white sheet out of it to signal for help.

4. Wait for the fire fighters to rescue you. Don't jump.

5. If the door was cool, open it carefully to see if there is smoke in the hall, or if there is heat pressure against it. Slam it shut if this is the case.

6. Make your way down the corridor or stairway with care.

7. On the way out of the building, sound the building fire alarms if there are any. Otherwise

pound on apartment doors and yell "Fire!" to alert your neighbors. Close all doors behind you to slow the spread of fire.

8. Once you are outside, notify the fire department if they aren't on the scene. (You should know where the nearest street alarm box and phone are located.)

Yes, it does pay to be prepared. A fire is a frightening experience, but the family that is prepared for such an emergency will know what to do. Practicing home fire drills is probably the most important preparation that will pay off.

This is posted next to elevators in a high-rise apartment complex in Boston.

## PRUDENTIAL CENTER APARTMENTS
## FIRE EMERGENCY RECOMMENDATIONS FOR OCCUPANTS

### FIRE IN YOUR APARTMENT

1) At anytime — Notify the Service Control Manager (lobby desk) through the intercom, or Niles' phone 536-9300. Tell him your apartment number and name. He will alert the Fire Dept. and all available building personnel. If the desk cannot be contacted immediately, call the Fire Dept. — telephone 536-1500 and give them the floor, apartment number and street address. As early as safety permits, contact lobby desk to advise of Fire Dept.'s notification.

2) Without further delay, leave your apartment, close the door, leaving it un-locked.

3) Alert other tenants on your floor, especially during sleeping hours.

4) Use the closest exit stairway.

5) Do not use the elevators.

6) Advise desk of any temporarily or permanently handicapped persons who might require assistance.

### FIRE OR SMOKE NEAR YOUR APARTMENT

1) At any time — Notify the Service Control Manager through the intercom or Niles' phone 536-9300. Tell him your apartment number and name and what you have seen. Do not assume that anyone has already called him.

2) Before trying to leave your apartment, place your hand on the door. If the door feels warm, do not attempt to open it as this indicates fire in the corridor.

3) If the door is not warm, carefully open it a small amount to check for smoke in the corridor.

4) If you feel that the corridor can be used, close your door, alert other tenants on your floor and use the closest exit stairway. Do not use the elevators.

5) If your apartment door is warm to the touch or there is heavy smoke in the corridor, keep the door closed. Seal places where smoke is entering, with wet towels. Contact Service Control Manager to tell him you are still within your apart-ment.

6) Fire extinguishers have been provided in each corridor Fire Cabinet. Instructions for their use have been mounted inside. These may be used by tenants and building personnel.

## TYPICAL CORRIDOR FLOOR PLAN SHOWING EXIT & FIRE EXTINGUISHER LOCATIONS

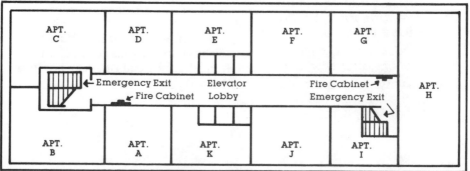

Chapter VI

# Ask What Your Fire Insurance Company Can Do For You!

## Keeping Covered

**Most people still believe that their home is their castle, and 98 percent of all Americans retain fire insurance to keep it that way.** Yet the castle can become a smoldering ruin and, in many cases, the insurance policy goes up in smoke along with everything else.

Frequently the homeowner has no idea where his coverage begins and ends. In a fire, the major investment of a lifetime can be taken in a matter of minutes. Would your insurance cover all fire damage? Would it pay your medical bills? Would all of your belongings be replaced?

Ninety-five percent of all Americans never read their insurance policies and have no idea of their coverage. And far too many people learn the extent of their coverage when it's too late — when a fire has already occured.

Though many homeowners are unaware of it, fire coverage of most homeowners' insurance policies has changed over recent years. More things can be done around the home to obtain a reduced rate. Certain parts of the policy are particularly worth careful reading.

### Check It Out

An inspection of your homeowners' policy is crucial and could reveal some interesting facts. You may need additional insurance or you may be paying for conditions that do not exist in your home.

Homeowners' policies vary widely among insurance companies. Fire coverage is included in all forms but the specifics of your particular policy should be looked at closely. The fire portion of the homeowners' policy has two main sections: property insurance, which covers loss or physical damage to your property; and liability insurance, which includes medical payments.

### Property Insurance

What is an adequate amount of insurance for you? The recommended amount of insurance coverage on a residence should cover at least 80 percent of the cost of replacing the house. If the construction replacement cost of your home is $50,000, the amount of insurance on it should be at least $40,000 — 80 percent of 50,000. It is important for you to note that even a partial loss, such as a fire that damages only one room, will not be fully covered unless the total amount of insurance on the dwelling at the time of the loss is 80 percent or more of the replacement cost.

Many people carry less coverage than 80 percent of the full replacement cost. Their property is insured on an actual cash value (replacement cost less depreciation). The homeowner with actual cash value insurance loses out to inflation, since appreciation in replacement cost has not been taken into account.

Simply stated: if an average six-room home had contents that at today's market would cost between $30,000 and $50,000 to replace and the items had been acquired over a ten-year period, they might be valued at an actual cash value of $15,000.

No homeowner wants to square off with inflation like this — and especially not after a fire. Yet one out of every two homeowners today are under-insured — they have never bothered to reevaluate their home in today's real estate market.

**Medical Payments**
In the event of a fire, the medical payments coverage contained in the liability section of your homeowners' policy will pay for medical attention to any visitor who was in your home during the fire. Since the basic amount of protection is only $500 for each person, you may want to purchase larger amounts. At the cost of medical treatment today, $500 can be used up very quickly.

80% Coverage

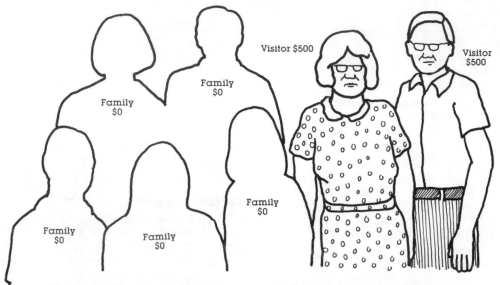

Visitor $500

Family
$0

Family
$0

Family
$0

Family
$0

Family
$0

Visitor
$500

These medical payments do not cover the homeowner or the members of the household. Additional health insurance is required to meet those needs; homeowners cannot assume that the fire section of their homeowners' policy will cover these medical costs.

Most homeowners' policies will also pay for your living in a hotel and eating at restaurants while your home is being repaired after a fire. However, this is not a full-fledged expense account. The insurance company will reimburse you up to the limits stated in your policy — but only for the difference between these expenses and your usual living expenses.

## Condominiums and Apartments

Condominium and apartment dwellers require a special type of fire protection insurance, and the insurance industry has tailored specific policies to those needs.

Usually, a condominium association buys the insurance covering the condominium's property — both the building and the structures. Condo owners buy policies to protect their personal property and liability.

Similarly, those who rent apartments can insure their personal belongings, while the landlord is responsible for insuring the entire structure.

But wherever you make your castle, comprehensive insurance coverage against fire is an absolute necessity.

## The Right Rate

Comparison shopping for the most protection at the best price is still the name of the game in the insurance market. The savings can be substantial. But there are several additional variables that will determine how much your coverage will cost.

Fire protection insurance takes into account diverse underwriting considerations. In addition to the age of your home and property, agents will look at your general neighborhood — its crime rate and if there's an arson problem, the distance from your home to the fire department, and even the width of the streets on your block. Ask your insurance agent what is looked for on these points.

You should also check that the agent doesn't assume too much about your home on the basis of its age. If your home has new fire-retardant shingles or improved insulation material, or even a new furnace, be sure to tell your agent — this could make a big difference in your premium.

By the same token, the installation of a new woodburning stove may increase the cost of your insurance, while the placement of ceramic tiles beneath the stove may prevent the cost from escalating. Carefully discuss these issues with your agent.

## How Rates Vary From One Company To Another in the Boston Area

| | Commercial Union | Allstate | Aetna Casualty | Hanover | Liberty Mutual |
|---|---|---|---|---|---|
| Back Bay | $803 | $585 | $904 | $747 | $658 |
| Dorchester | 865 | 614 | 961 | 806 | 710 |
| Winthrop/ Chelsea | 808 | 588 | 762 | 752 | 626 |
| Brookline | 716 | 596 | 688 | 665 | 552 |
| Plymouth (and most of Plymouth county) | 578 | 427 | 456 | 515 | 451 |
| Worcester | 566 | 411 | 513 | 496 | 464 |

*These are the annual premiums for a standard $75,000 homeowners policy ($100 deductible) for a wood-frame house in the towns shown. The premiums do not reflect any discounts or special circumstances.

**Source:** Insurance Services Office. May 1981.

## Going After The Discount

To qualify for a lower premium, homeowners are sometimes required to make changes in their homes. Though these changes may require some money up front, they are home improvements that should increase the value of your home and provide substantial savings over the following years.

Smoke detectors are a major factor in obtaining these discounts, but most insurance companies are also asking homeowners to install several precautionary devices before offering reduced premiums. Some insurance companies are offering a 5-percent discount to homeowners who install dead-bolt locks on all outside doors, smoke alarms on each floor of the residence, and fire extinguishers. The same companies often offer an additional 10-percent discount if there is a central fire/burglar alarm system. Comparison shopping will reveal who offers the best buy in your area.

## Preparing Your Home

The standard homeowners' policy covers your furniture and belongings up to half of the amount for which you insure your house. If you buy a $50,000 policy, your belongings are insured up to $25,000.

On some items, like television sets, the company will take the age of the property into consideration, and subtract something for depreciation.

The more proof you have about what you own and what it's worth, the better. A detailed inventory of household furnishings and personal belongings can be just as critical as the insurance policy itself in the event of a fire. Most people have a great deal of difficulty recalling exactly what they own despite their familiarity with the objects.

An inventory establishes the purchase date and cost of every major item. It identifies exactly what was lost and aids the settlement of the insurance claim. It also verifies your losses for income tax deductions. And it is more useful than anything else in determining the value of your personal belongings and your individual insurance needs. Without a detailed inventory, the homeowner who just suffered a fire is really at a loss.

## A Picture Is Worth a Thousand Words

Many insurance companies have developed special forms to assist you in taking inventory. To back up the written inventory, photographs should be taken of each wall of each room with closets or cabinet doors open. On the back of each picture, the date should be written, as well as the general location and the contents shown.

Several new companies have sprung up across the country who are in the business of taking photographic inventories. Most charge a modest fee to provide both the insurance company and the homeowner with an extensive inventory plus photographs.

Both personal and business property are usually

photographed with a wide-angle camera lens to show existing property. Closeup shots of prized possessions are also provided, including valuable collections, antiques, paintings, silverware, china, and jewelry.

Some photographic inventory companies store the original negatives of the photos as well as any required written documentation in a vault for ten years after completion of the inventory. Others take only the photographs, and it's up to the homeowner to have the film developed and complete documentation. Still others are now involved in the recording of slides and videotapes of personal property.

However photographs of the property are much easier to show to an insurance claims adjuster than slides or videotapes. While the slides may reduce the cost of your photographic inventory, videotapes may prove to be quite costly. Photographic inventory packages can be tailored to meet individual homeowners' needs, and so it is advisable to take advantage of the free estimates and consultation services of these companies.

In considering the cost of the photographic inventory services, it is a good idea to consider the cost of doing a comparable job yourself. Most services provide the homeowner with 8" x 10" prints, and the average photographic inventory will include 20 prints. Since the average cost of having prints of that size developed yourself is $4.75 each, developing alone would cost $95. For about that price or a little more, you can have the entire job performed professionally and recieve the written documentation as well.

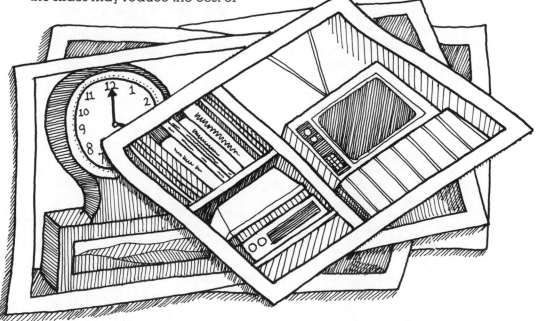

The photographic inventory is not only critical in the event of fire but also provides protection against theft. Experts report that the completed photographic inventory aids police in identification of stolen merchandise, and sometimes in the recovery of the goods and conviction of the criminal.

Finally, if you own anything that is particularly valuable — a painting, silverware, furs, antiques, jewelry — a photograph is not enough. Have a written copy of the appraised value.

## Insurance Protection Checklist

| Yes | No | |
|---|---|---|
| | | Does your insurance cover at least 80 percent of the cost of replacing your house? |
| | | If there is a visitor in your home who is injured during a fire, do the medical payments coverage in the liability section of your homeowners' policy give adequate compensation? If it is only $500 per person you should consider purchasing more. |
| | | Does your policy cover adequate living expenses while your home is being repaired? |
| | | If you live in an apartment, do you have adequate coverage of your personal property? |
| | | Have you shopped around for the best rate for your insurance policy? |
| | | Have you advised your agent that you have smoke detectors in your residence, and that you qualify, therefore, for a discount? |
| | | Do you have a detailed inventory of household furnishings and personal belongings? |
| | | Do you have a photographic record to back up the written inventory? |
| | | Do you have an appraisal of particularly valuable items to back up the written inventory and the photographic inventory? |

If you have answered Yes to all these questions, you are keeping covered.

# Chapter VII

# Just In Case...First Aid

Emergency Treatment of Injuries

**A lot of people take a "head in the sand" attitude about medical treatment of the injured.** Yet it's extremely important to be able to react properly to someone who has been injured. A few simple procedures until a doctor can take over can sometimes save a life.

### Basic Procedures for First Aid
Do not move the injured person unless it is necessary because the location is a dangerous one. In all of the procedures try not to move the person any more than you must.

There are several conditions that can be considered life-threatening, but respiratory arrest and severe bleeding require attention first. Once they have been alleviated, attention can be focused on other obvious injuries.

Keep these general guidelines in mind:

Keep the injured person lying down in a comfortable position while you try to discover how serious the injuries are.

Cover the injured person to prevent chills.

Watch for hemorrhage, stoppage of breathing, wounds, fractures and dislocations.

Remove enough clothing to determine the extent of injury. Rather than attempt to remove clothes in the usual manner, thus subjecting the injured person to a great deal of movement, rip seams to get them off.

**First in importance is to control bleeding.** If the wound is small, press on it with a piece of sterile gauze. A tourniquet may be needed if there is serious bleeding, but they are dangerous and should not be used if bleeding can be checked readily otherwise.

**If breathing stops, apply mouth-to-mouth or mouth-to-nose artificial respiration. (See page 113.)**

Never give water or other liquids to an unconscious person.

Remain nearby until qualified medical personnel arrive.

### Artificial Respiration
Here are the basic steps to take in giving artificial respiration. It is the foremost method recommended today by leading authorities including the American Red Cross. Following are the procedures for the mouth-to-mouth technique:

1. Make sure air passages are not blocked. With the person lying on his back, remove foreign matter lodged in the mouth or throat. To do this, turn his head to one side and clear it out with your fingers or a piece of cloth.

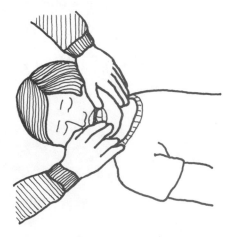

2. Kneel close to the person's head and tilt the head back so that his chin is pointing upward.

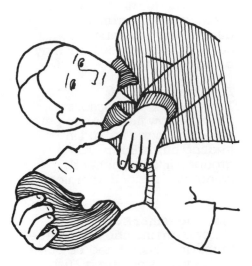

3. Listen closely for the sound of breathing from the victim's mouth and nose. Watch for a rise and fall of the chest. This should inform you if the victim is still breathing.

4. If there is no sign of breathing, use the thumb of one hand to open the person's mouth and grasp the lower jaw (quickly wrap your thumb first, for protection). Lift the jaw upward and forward so that it juts out. Pinch shut the victim's nose with your other hand. Breath deeply, place your mouth over his and blow into the victim's mouth.

5. Again listen and look for signs of respiration. Watch the chest, and when it expands, take your mouth off his and let him breath out. Repeat this procedure twelve to twenty breaths a minute.

6. If the person fails to start breathing, blow again into his mouth. Continue this procedure until respiration revives.

7. Watch the person carefully after he starts to breath again for further breathing difficulty.

8. Arrange for an ambulance for the injured person to be taken to a hospital.

**\*Note:** If this technique is used on a very small child, follow steps 1, 2, and 3. Then place your mouth over the baby's mouth and nose, closely so that air cannot get out, and blow in. Blow gently, up to twenty breaths a minute.

Now let's go over the injuries most frequently caused by fire.

## Burns

The key to giving proper first aid for a burn lies in recognizing the severity of the burn.

The severity of a burn depends on its size, the depth of penetration into the skin, and the kind of person burned.

The larger the body area affected, the more serious you should consider the burn. A general rule is that burns covering 10 percent or more of a body require immediate hospitalization.

The severity also depends on the part of the body involved. Burns on critical areas such as hands, feet, face, and genitals are extremely serious.

The "who" of a serious burn situation includes infants, children, and the elderly and sick. They are most likely to develop complications.

The severity of a burn can also vary depending on the amount of time the individual is actually exposed to the heat source. The longer the exposure, the graver the injury.

## Types of Burns

Burns are classified as one of three basic types: first-degree, second-degree, and third-degree.

A **first-degree burn** is characterized by mild swelling and a reddish color. There is usually no blistering. First-degree burns can be caused by scalding or by brief contact with very hot substances.

A **second-degree burn** is usually marked by blisters, and can be quite painful. The surface of the burned skin may be wet.

A **third-degree burn** exhibits a charred appearance. The affected area may be brown in color. A third-degree burn may be painless at first.

## Treatment Of Burns

The following two sections describe how the average person can apply basic first aid in treating burns. Burns will be grouped into two general categories, minor and major burns.

## Treating Minor Burns

Minor burns include first-degree burns and second-degree burns that cover a small area. Such burns are not life-threatening.

Immersing minor burns in cool, clean water for about five minutes will help to lessen pain.

Pain can also be reduced by applying an insulated cold pack for no more than fifteen minutes. Do not immerse burns of any kind in ice water. It is also a sound practice to gently dry the area after soaking and cover with a dry sterile bandage. It is permissible to spread oil or lotion over mild burns. do not use absorbent cotton directly on a burn because it sticks and is difficult to get off.

### Treating Major Burns

Major burns include sizeable second-degree burns and third-degree burns of any size. Major burns can put a victim's life in jeopardy, so certain measures must be taken immediately.

Basically, for large second-degree burns, immerse the burn in cool water until the pain lessens. Gently pat dry and cover with a clean cloth. Do not apply ointment. The blisters that characterize second-degree burns should not be broken or cut away. The patient should be closely watched for signs of shock.

Unlike first- and second-degree burns, third-degree burns should not be soaked. Keep burned arms and legs elevated if possible. Be wary of removing the victim's clothes. Charred skin may be stuck to the fabric. A good practice is to leave clothes on unless the clothes are smoldering.

The best thing that can be done for victims of third-degree burns is to cover them with a clean, dry sheet and try to keep them quiet and motionless.

It is crucial to get severely burned people to the hospital as soon as possible. Only a medical center or burn center will have the modern burn-treatment facilities essential for swift recovery.

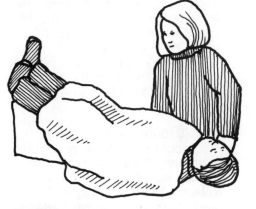

### Eye Burns

For superficial burns in and around the eye, flood the affected eye with water. Cover with a dry, antiseptic wrapping and bandage the dressing place.

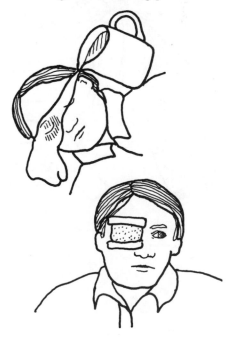

Admonish the victim about rubbing the afflicted eye. In the event of chemical burns to an eye, take care that the chemical does not drain into the other eye when flushing with water. Also, take care to remove contact lenses before flushing.

With eye burns, as with burns to any crucial part of the body, it is essential to contact a physcian immediately.

### Electrical Burns
Electrical burns can vary widely in severity. Burns from low-voltage equipment might be rather mild. High-voltage electrical burns are very serious and must be treated at a hospial or special burn center.For either type, call for outside medical assistance right away.

In the case of electrical accidents, which happen frequently in home fires, it may be necessary to rescue a victim entangled by electrical wires. The electrical current must be turned off before an attempt is made to reach the trapped person. Disconnect the electricity by tripping the circuit breaker or by pulling the plug on the electrical device. If you cannot shut off the power, push the victim away from the live wires with a non-conductor such as a long dry wooden board.

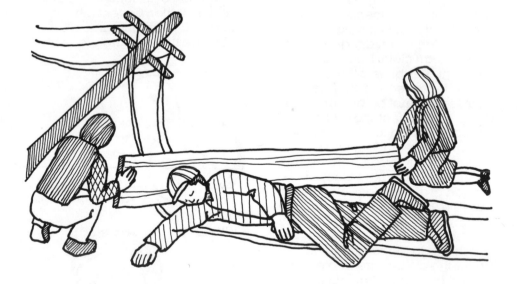

## Chemical Burns

Burns from chemicals commonly used around the house or at work can cause severe burns. Prompt action is needed to stop chemicals from penetrating further into the skin. Generally it is a good practice to remove all clothing from the burn and to soak chemically burned skin in lukewarm water. Then cover the burn with a sterile dressing.

Penetration of chemicals deep into the skin can result in severe complications, such as an allergic reaction. Contact the proper medical authorities (such as a poison control center) immediately.

## Infections

The breaks in the skin caused by burns can lead to infection. Infection can occur when bacteria enter body tissues through wounds.

A victim may suffer a serious infection only a few hours after being injured. The infection can lead to serious complications if it is not quickly treated.

The following is a list of some of the symptoms of infection:

- redness
- swelling
- pus
- constant pain
- feeling of heat around the wound

If evidence of infection is detected, you should have the victim lie down. Remove any foreign bodies (dirt, glass, sand) from the wound. Pay particular care to keeping the affected area still. Once sterile bandages are in place, do nothing to remove them.

These are temporary measures only. Again, make contact with emergency personnel without delay.

## Shock

Shock is a very serious condition caused by many types of severe injuries. In physical terms shock is a lowered state of body functions such as blood flow. Brain damage can result from shock when the victim's brain cannot get enough oxygen.

The following is a partial list of the symptoms of shock:

- the victim is restless or drowsy
- the victim is thirsty
- skin that is moist and clammy. Also, skin that has lost color
- breathing that is rapid and weak
- eye pupils that are dilated.
- nausea.

There are several steps in the treatment of people affected by shock.

First, the victim, in most cases, should be kept prone. Lying down may help better blood circulation. Prop up the victim's legs about a foot off the ground. This may help circulation too. If the victim is bleeding, try to stop the blood flow.

Second, the victim should be kept warm. This is essential because of the depressed condition of the body functions. Put a blanket over the victim. Don't get carried away, however: providing too much heat to the person in shock can cause serious damage.

These two procedures can be administered immediately. Another treatment of shock victims should be employed only when the arrival of rescue squads is not expected for hours. The third treament is feeding fluids to the victim. The preferred fluid is a mixture of water, salt, and baking soda. (For each quart of water, add a ½ teaspoon of baking soda and 1 teaspoon of salt.) The water should be around body temperature. Never give the victim alcohol.

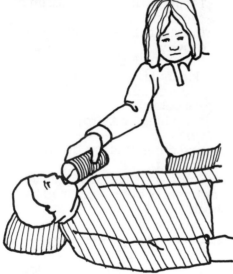

## Smoke Inhalation

Smoke inhalation is one of the serious dangers in any fire. Breathing in smoke or toxic gases can cause the lung passages to swell, obstructing the flow of vital air.

If you know that a person has encountered heavy smoke in a burning building, or if you notice a person has burn marks around the nose and mouth, you should assume that person is a victim of smoke inhalation and take appropriate action.

The first thing to look for is the pattern of the injured persons breathing. If the person is breathing, keep his air passages clear. Remove foreign matter like bubble gum from the mouth. Keep the person lying down, unless he expresses a desire to sit. He may be allowed to sit up if this helps his breathing. Contact medical personnel immediately, as this is an extremely serious injury.

## First Aid Supplies for the Family Medicine Chest

|  |  |
|---|---|
| ✔ | Every household should have a first aid kit or cabinet. Keep one in your automobile as well. Supplies should include: |
| ☐ | Adhesive tape of various widths |
| ☐ | Sterile cotton |
| ☐ | Sterile bandages |
| ☐ | Small ready made adhesives with tiny sterilized gauze patches |
| ☐ | Antiseptic soap |
| ☐ | Mercurochrome |
| ☐ | Iodine |
| ☐ | Wash basin for soaking |
| ☐ | Ointment to spread over a burn |
| ☐ | Scissors reserved for medical use |

# Chapter VIII

# America's Burning Question

## Your Answer As A Citizen

## The Nation's Fire Problem

America, the richest and most technologically advanced nation in the world, is tragically ahead of all the major industrialized countries in per capita deaths and property loss from fire. Even taking into account the differences in reporting procedures which make international comparisons somewhat unreliable, fire death statistics show the United States as the leader, based on estimates of the National Fire Prevention and Control Administration. Statistically, it is estimated that more than 300 destructive fires rage somewhere in our country each hour. Two-thirds of the people who die in them will be in their own homes. These fires will have caused more than $300,000 worth of property loss, at least one death, and thirty-four injuries. The crippling, disfigurement, scars and terrifying memories that remain with the thousands of Americans who are injured by fire every year cannot be calculated. Fire is one of the greatest wasters of our natural resources.

Although the young, the old, and the helpless are the most susceptible, our nation's fire fighters are among those who suffer the most from the prevalence of fires. They have the most hazardous profession in the world. Many of them, especially in the smaller cities, do not have adequate training. But even under the best of circumstances, every time a firefighter goes to a fire in the line of duty, the risks are tremendous.

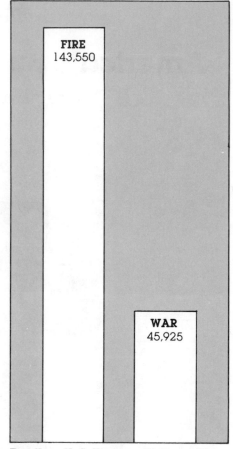

**Deaths—U. S. Fires vs. Vietnam War**
Comparisons of deaths in U.S. military personnel (Army, Navy, Coast Guard, Marine Corps, and Air Force) resulting from actions by hostile forces in Vietnam, 1961 through 1972, and deaths from U.S. fires for the same period. (Statistics from the Department of Defense and the National Fire Protection Association.)

## Causes:

What is the reason for this poor record in fighting fires? *America Burning*, the report of the National Commission on Fire Prevention and Control, found the following:

### Fire's Low Priority as a Hazard

Ignorance and indifference, say individuals and organizations in the fire protection field, are the two largest contributors to the magnitude of the fire problem in the United States. In some cases, investment in fire protection is given low precedence due to lack of understanding of fire's threat. While anyone who has survived a fire will never forget its destructive potential, for most Americans fire seems to be a remote danger that justifies indifference. In others, the fire services are unaware of the technological state-of-the-art in their field. Or, there are the fire department administrators who pay lip service to fire prevention but actually do little to promote it. In some communities the public shares this indifference. Citizens tend to see the fire department as an army of heroic fighters who stand ready to rescue people and suppress fires, but fire prevention is often not seen as the fire department's role by either citizens or the fire fighters.

### Inadequate Building Codes

Building designers, as a rule, feel that it is sufficient to satisfy the minimal safety standards of the local building code. Frequently, both they and their clients assume that the codes provide completely adequate measures, which is not necessarily true. In some instances, they view fire as something that will never happen to them. In others, they prefer to pretend that there is no risk of fire because of the cost of fire prevention measures, or they may occasionally opt to take the risk balanced by the provisions of a fire insurance policy.

### Insufficient Federal Government Involvement

The influence of governmental agencies in the field of fire prevention has been increasing, but it is significant that the two countries that lead the world in fires, the United States and Canada, have less federal governmental agency participation, except at the municipal level, than have other countries. The Occupational Safety and Health Act of 1970, has given the Department of Labor fire prevention powers which concerns all employees, businesses, industries, institutions, farms, and other places of employment that engage in interstate commerce. The Department of Transportation and the Department of the Army and Navy have fire prevention efforts and training. Traditionally, the Department of Agriculture has had the greatest impact on the public, mainly through the Forest Service. The National Bureau of Standards and the National Fire Prevention and Control Administration, under the Department of Commerce have made many contributions in the field. Nevertheless, although some of the Federal programs that do exist in the United States, are excellent, they affect only a part of the total fire problem.

## Public Apathy

Americans today are very concerned with air pollution, which is less destructive, but generally indifferent to and ignorant of the heavy toll that is taken by fire. In spite of many efforts, the problem has not yet reached the consciousness of the American public with the same impact that led them to reach into their pockets and support the fight against poliomyelitis so that it is virtually eliminated today.

This indifference to destructive fire as a national problem is reflected in the fact that there is not an all-out war against arson and false alarms. Professionals such as fire officers, arson and insurance investigators, who examine the evidence in the ashes, believe that at least half of all fires of unknown origins are arson cases. Two types of arson are increasing: the fraudulent fire for insurance profit, and fires set for revenge. By definition, arson is the non-accidental, deliberately set fire, for whatever reason: profit, passion, or madness.

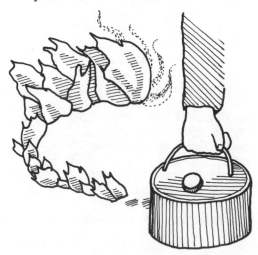

## Changes In American Society

Partly because of this national indifference, and partly because of rapid changes in American society that have added new problems to the difficulties in handling the fire problem, our methods of handling the fire problem are not adequate to meet the needs of today. Anthropoligists refer to this as a "culture lag." Our methods of handling the fire problem are tuned to an America of yesteryear, and are inadequate for our contemporary, much less, future requirements.

One of the important differences in our modern society as compared to the way it was half a century ago, is the increasing urbanization of the United States. Fifty years ago, about half our population lived in cities, whereas now about three-fourths of the population are city-dwellers. Even though distances from firehouse to fire site are generally shorter in urban areas than in rural areas, crowded city streets slow the progress of the fire trucks as they attempt to reach a fire. Minutes lost can make a serious difference in the effectiveness of fire response. High-rise buildings are the bane of fire fighters existence. Automobiles parked near buildings make it difficult for the ladders to reach beyond seven or eight floors, and the ladders for most large city fire trucks don't reach beyond the ninth floor even when fully extended. In addition, it is difficult to vent heat and smoke in modern air-conditioned buildings.

warmed with dangerous portable or make-shift heaters. Furthermore, these neighborhoods are natural breeding grounds for discontent. For fire departments, the problem erupts in riots to be controlled, fires that were deliberately set, false alarms, and even harassment of fire fighters.

As fire fighters have become aware of the benefits of unionization, they have seen that they have been bypassed while other municipal departments have received increased salaries and benefits. This has resulted in job actions — slowdowns, massive sick leaves, and strikes — jeopordizing community fire protection.

At the same time, city governments have been confronted with severe financial problems, causing them to decrease the size of their fire department staffs, and demands that they operate more efficiently without sacrificing the public's safety from fire. Genuine

Crowded ghetto areas are another big city fire problem. Fire hazards are created by the crowding of poor families in tenement buildings where the upkeep of the properties is badly neglected. Frequently the central heating, if there is any, is not working, and tenants' rooms are being

economic problems often stand in the way of deeper investment in fire protection, especially in these times of recession. In Boston, recent cutbacks have caused layoffs of personnel, and curtailment of fire prevention programs in the schools and community.

Possibly the most important change affecting our country's fire protection is the technological revolution. New materials and new products fill our man-made environment and little is known concerning their hazard potential when they burn. New chemicals and other hazardous materials are being produced, shipped, and used around the country. Fire departments often have no experience or knowledge of how to deal with the toxic gases they produce when burning. Technological advances in the fire fighting field have been relatively slow when compared to such fields as aviation, large-scale construction, and electronics.

**Recommendations of the NFPA**
The 1973 report of the National Fire Commission on Fire Prevention and Control recommended the establishment of a U.S. Fire Administration to provide a national focus for the nation's fire

problem and to promote a comprehensive program with adequate funding to reduce life and property loss from fire. This was to be a supplementary agency for the sake of a more coherent effort to reduce fire losses in our country. It was recommended that:

— There should be more emphasis on fire prevention. Fire departments need to expend more effort to educate children on fire safety, to educate adults through residential inspections, to enforce fire prevention codes, and to see that fire safety is designed into buildings. These efforts should be continuously evaluated to see which measures are most effective in reducing the incidence and destructiveness of fire.

— The fire services need better training and education. The training for fire fighters and officers ranges from execellent as in some large cities, to almost non-existent, as in many rural areas. Better training would improve the effectiveness of fire departments and reduce fire fighter injuries. Better education provides the key to developing leadership for fire prevention.

— Americans must be educated about fire safety. Most destructive fires are caused by the careless actions of people, largely through lack of concern and ignorance of hazards. Many fires caused by faulty equipment rather than carelessness could be prevented if

people were trained to spot the faults before it's too late. And many injuries and deaths could be prevented, if people knew how to react to a fire, whatever its cause.

— The design and materials in the environment in which Americans live and work present unnecessary hazards. The emphasis should be on a labeling system (to be developed by the Consumer Product Safety Commission) for materials and products, so that consumers, at the time of purchase, know what risks are involved. The impact of new materials, systems, and buildings, on users and the community should be assessed during design stages, well before use. Careful analysis and filing of a fire safety effectiveness statement should permit recognition of faults before tragedy strikes.

— The fire protection features of building need to be improved. There is a need for automatic fire extinguishing systems in every high-rise building and every low-rise building in which many people congregate. Economic incentives for built-in protection should be provided. Perhaps most important, Americans need to be encouraged to install early-warning fire detectors in their homes where most fire deaths occur.

— Important areas of research are being neglected. The state-of-the-art in fire fighting, in treatment of burn and smoke victims, in protecting the built environment from combustion hazards, points to the need for a major expansion of research and development in these areas. Progress in most of these areas is hindered by a lack of fundamental understanding of the behavior of fire and its combustion products.

**What You Can Do To Help**
In our society, it is an acceptable principle that when voluntary safeguards are inadequate, the government ought to intervene to protect the citizens. There are unscrupulous merchants, so there are laws and court procedures to protect consumers. There are additives used by food processors which may be injurious to health, so the government needs to test these chemicals and ban them if appropriate. If there are drivers who drink alcoholic beverages, then there is a need for government efforts to keep them off the highways.

Still, we do not want government regulating every aspect of our lives, and wherever possible, we prefer to avoid government regulation, and use it only as a last resort. As private citizens in America, we have an obligation to take steps to protect ourselves and others from fire.

In this book, we have tried to make clear that fire is a potential threat to the life and well-being of everyone — no one is immune to harm from fire. We have pointed out areas in daily living where good judgment can minimize the chance of fire and make the difference between life and death if fire strikes. At the very least, the precautions to be taken in the home are well-established:

- A well-maintained heating system.
- No overloaded electrical circuits.
- Inflammable liquids stored in tightly fitting containers and away from heaters and furnaces.
- Absence of rubbish.
- Unobstructed stairways.
- Matches out of reach of children.

Beyond these minimal precautions are these positive steps:

- Most important — the installation of early-warning smoke detectors.
- The installation of fire extinguishers, and where possible, of escape ladders.
- The family discussion and rehearsal of steps to be taken during various kinds of fire emergencies.

Away from home, be alert whether at work or play:

- If there appear to be dangerous conditions at the place of work, report them to the Occupational Safety and Health Administration.
- Make it a habit to note the location of fire exits when entering a building or a restaurant.
- When you see them, call your neighbor's attention to potential fire hazards such as unattended leaves burning.
- Be aware of elderly and handicapped people in your neighborhood, and give them

any assistance they might need such as removing trash from the yard. Be sure they have a means of summoning help and escaping from fire.

- Acquaint friends with the subject of fire safety.
- Initiate fire prevention programs in your communtiy, such as a community smoke detector campaign, if they don't already have them.
- Participate in the programs to educate and inform citizens of the threat of fire, and the proper reaction in case of fire.

**The American Dream**

Fire is both friend and foe. We can't live without the major sources of destructive fire — electrical energy, lumber, petroleum and its distillates — so we must learn to live in harmony with them.

When the United States began, resources were so abundant that whatever was consumed could be easily replaced. During those years, as America matured, death was accepted stoically, whether by typhoid, or fire, even among young children. But with advances in medicine, expectations changed. As more diseases were conquered, people stayed healthier and lived longer, and dying was accepted only for the aged and infirm.

In recent years, Americans have begun to realize what other nations have long known: resources are limited and need to be carefully managed. Attention is being given to ecological considerations because of the education of the public to the need.

It is to be hoped that this same attention and education will be given to America's burning problem in future years. The Japanese, who for centuries built their homes of extremely inflammable materials consider a destructive fire a disgrace to the person who allows it to happen; at one time the punishment was crucifixion. A Japanese proverb translates: "There is no one who fails to get excited when the neighbor's house is on fire." In other words, other people's troubles do not interest most of us until their problem comes close to home. Then we are willing to do something.

The threat of fire must be brought "close to home" for Americans. They need to be educated to the dangerous enemy they have in destructive fire, and take the view that a fire caused by one American is a danger to all and an unfair cost to our fellow citizens.

So, the new American dream is to have instilled fire consciousness in every citizen. Then the day could be envisioned when every American home will have its own automatic fire department: early-warning smoke and fire alarms that activate automatically the extinguishers to put out the fire. Then thousands of lives would be saved every year, millions of dollars of our nation's resources would be saved from ruin, hospitals could be emptied of beds for burn and smoke injury patients, fire departments could be pared to a fraction of their present size, and fire insurance would be as inexpensive as marriage licenses.

Your answer to America's burning question is to help make this dream a reality.

## FIRE'S DO'S AND DONT'S

Educational materials distributed by the National Fire Protection Association, the National Safety Council, the American Insurance Association, and others emphasize the major gaps in everyday knowledge and practice:

### Before the Fire Starts

- Remove trash and stored items of outlived usefulness, particularly from the vicinity of furnaces and heaters and from hallways and exit areas.

- Exercise care in the use of electricity. Do not overload electrical outlets with many appliances, use only appropriate fuses, and do not hang electrical cords over nails or run under carpets. Have cords replaced when they begin to fray or crack, and have electrical work done by competent electricians.

- Do not store gasoline or flammable cleaners in glass containers, which can break, and avoid storing them inside the home. Do not keep more flammable liquids on hand than you really need. To avoid the danger of spontaneous ignition, dispose of rags wet with oil, polishes, or other flammable liquids in outdoor garbage cans.

- Inspect your home and workplace often for these and other hazards.

- Sleep with bedroom doors closed. In the event of a fire, you will gain precious minutes to escape.

- Learn how to extinguish common fires in early stages the best way. Roll a person whose clothing is on fire; use a proper portable extinguisher or even a handful of baking soda to extinguish a fire on your stove.

- Clothing afire is a prelude to tragedy. Buy garments, such as children's sleepwear, that meet Federal flammability standards as they become available. Do not wear (or permit children to wear) loose, frilly garments if there is any chance at all of accidental contact with a stove burner or other source of fire.

- Exercise extreme care with smoking materials and matches, major causes of destructive fire. Do not leave these where children can reach them.

- Invest in fire extinguishers, escape ladders, and — most important — early warning (smoke or products-of-combustion) fire detector and alarm devices.

### After a Fire Starts

- If you see, smell or hear any hint of fire, evacuate the family immediately, but don't compound tragedy by attempting a rescue through a gauntlet of flames or thick smoke. Call the fire department as soon as possible. Don't attempt to extinguish a fire unless it is confined to a small area and your extinguishing equipment is equal to the task.

- If your clothing ignites, roll over and over on the ground or the floor. Running will just fan the flames. Teach the proper procedure to your children.

- Before opening your door when you suspect fire in another part of the building — as in a hotel, for example — feel the inside of the door with the palm of your hand. If it's hot, don't open it. Summon aid, if possible, and go to a window and await rescue. If smoke is pouring into the room under the door, stuff bedding or clothing into the crack.

- In smoke, keep low. Gases, smoke, and air heated by fire rise, and the safest area is at the floor. Cover mouth and nose with a damp cloth, if possible. Don't assume that clear air in a fire situation is safe. It could contain carbon monoxide, which, before it kills you affects judgment, hampering escape.

Cut out and post as a reminder.

# Appendix I

After interviewing several hundred adults and children around the United States, the following questionnaire was devised by the National Commission on Fire Prevention and Control. The questionnaire is now being used in schools, together with an answer sheet, so that students can learn, while correcting their own papers. Should you wish to test your fire safety knowledge, the answers are to be found on page 134.

## FIRE SAFETY QUESTIONNAIRE

Student ☐   Fire Safety Teacher ☐   Age ___ Schooling: Public ☐ Private ☐

Teacher: ☐   Previous Fire Training, _____
                                      Where (if any) school, Scouts, Army, industry, etc.

Address: _____
              City                                    State

Sex: Male ☐   Female ☐ _____

1. If your house began to fill up with thick, black smoke, what would you do? (answer fully)

   _____

   _____

2. What would you do if you woke up at night, smelled smoke, and found that your bedroom door was shut, but hot when you touched it?

   _____

   _____

3. Will the clothing you have on now burn?

   _____

4. What would you do right now if your clothing caught on fire?

   _____

   _____

5. If you were trapped in a bedroom on the fifth floor with flames outside in the hall and smoke pouring in under the door (with no telephone and no fire escape), what would you do?

_____

_____

6. (a) When you go to a strange place (movie house, friend's house for the night, hotel, restaurant, etc.), do you check to see where the exits or fire escapes are?

_____

(b) If the answer to 6(a) was "Yes," do you depend on being able to see the exit to find it, or do you figure out how to find it in the dark or in thick smoke?

_____

7. Do you have a family escape plan, including ways of getting out of your house if the stairs or doors are blocked by fire, **and a meeting place** outside the house?

_____

_____

8. What should you do (or your wife or mother do) if the frying pan catches on fire?

_____

_____

9. Carbon monoxide is produced by almost all fires. What effects does it have on you before it makes you sleepy and kills you?

_____

_____

10. Assume you plan to hang by your hands from a window ledge and then drop to the earth below. Estimate in feet the distance you could drop and still have a 50:50 chance of surviving without serious injury.

_____

11. (a) What is the reason for having fuses in an electric circuit?

_____

_____

(b) What strength fuse should be used in an ordinary lighting circuit?

_____

12. What number should you dial to report a fire by telephone, and how should you report it?

_____

_____

13. When is an electric cord dangerous? (give at least two examples)

_____

_____

14. When is a plug dangerous?

_____

_____

15. What should you do if you discover a large fire in your basement?

_____

_____

16. If you are trying to light a gas oven or burner and the first match goes out too soon, what should you do?

_____

_____

17. What is meant by "spontaneous combustion" or "spontaneous ignition"?

_____

_____

18. How should you store oily or greasy rags?

_____

_____

19. Why should gasoline be stored only in metal cans with self-closing caps?

_____

_____

20. Should you put out an electric fire with water?

_____

## SELF-SCORING THE FIRE SAFETY QUESTIONNAIRE

| Questions | Safety Score (points) |
|---|---|
| **Question 1.** If your house began to fill up with thick, black smoke, what would you do? (answer fully) | |
| If your answer included getting beneath the smoke by crouching or crawling (to evade harmful combustion products), give yourself | 3 |
| If your answer included getting out of the house, give yourself | 3 |
| If your answer included rousing the rest of the household, give yourself | 3 |
| If your answer included calling the fire department, give yourself | 3 |
| If your answer included opening the windows without first closing doors (to keep the air from the fire) subtract 3 points. | |
| **Question 2.** What would you do if you woke up at night, smelled smoke, and found that your bedroom door was shut, but hot when you touched it? | |
| If your answer did not include opening the hot door (which would expose you to killing heat), give yourself | 4 |
| If your answer included calling for help by phone or from a window, or finding an alternative way out, give yourself | 3 |
| **Question 3.** Will the clothing you have on now burn? | |
| If your answer is yes, give yourself<br>**Note.** — It is hoped that in the future this question will have to be deleted, as flame resistant materials become more available. | 3 |
| **Question 4.** What would you do right now if your clothes caught on fire? | |

If your answer included dropping and rolling (to extinguish the flames by smothering them) give yourself                                      **3**

If your answer included running (which fans the flames) subtract 3 points.

If your answer included going to draw water (which takes too long) subtract 3 points.

If your answer included wrapping up in a blanket, coat, or rug, but remaining vertical (thus permitting continued inhalation of smoke), give yourself only                      **1**

---

**Question 5.** If you were trapped in a bedroom on the fifth floor with flames outside in the hall and smoke pouring in under the door (with no telephone and no fire escape), what would you do?

If your answer included stuffing something into the offending crack to reduce the smoke entering the room, give yourself                                      **3**

If your answer included yelling from the window for help, or hanging something out the window to attract fire fighters' attention, give yourself               **3**

If your answer included jumping, subtract 3 points.

If your answer included opening the window a crack, top and bottom to vent the smoke and you did not leave a door open, so air could reach and fan the fire, give yourself                                      **2**

If your answer included finding better air by keeping low or breathing air from outside the window, give yourself      **2**

If your answer included making a rope out of bedsheets, curtains, etc., give yourself                                      **1**

If you said you would make it, but not use it unless forced to, give yourself an additional                              **1**

**Question 6 (a)** When you go to a strange place (movie house, friend's house for the night, hotel, restaurant, etc.), do you check to see where the exits or fire escapes are?

If you habitually check the exits when you stay at hotels, inns, motels, etc., give yourself     1

If you check to see where the exits are when at a restaurant or staying overnight at a friends house, give yourself     1

---

**Question 6 (b)** If the answer to 6(a) was yes, do you depend on being able to see the exit to find it, or do you figure out how to find it in the dark or thick smoke?

If your answer to 6(a) was no, give yourself no points for question 6(b)

If your answer to 6(a) was yes, and you do not rely on being able to see the exit signs, but figure out how to find an exit in the dark in thick smoke, give yourself     1

---

**Question 7.** Do you have a family escape plan (including ways of getting out of your house if the stairs or doors are blocked by fire), and a meeting place outside the house?

If you have a way out of your house, if the stairs and doors are blocked by smoke give yourself     2

If you have a planned place to meet outside the house which the whole family knows about, give yourself     2

---

**Question 8.** What should you do (or should your wife or mother do) if the frying pan catches on fire?

If your answer is to smother the fire with the lid or baking soda or to use a dry powder (all purpose) or $CO_2$ fire extinguisher, give yourself     3

If your answer is to smother the fire with salt or a wet towel, give yourself     2
(Sand and dirt are acceptable answers if cooking outside.)

If you threw water on the fire or used a soda-acid fire extinguisher or a water-pump tank type of extinguisher (water may spread the fire over the kitchen), subtract 3 points.

If you attempted to carry the flaming frying pan, which may ignite your clothing, spill, or become too hot to hold, subtract 3 points.

If you threw flour, which explodes, at the fire, subtract 3 points.

---

**Question 9.** Carbon monoxide is produced by almost all fires. What effect does it have on you before it makes you sleepy and kills you?

If your answer reported that carbon monoxide has no effect, or that it makes you cough, your eyes water, or smells badly, subtract 2 points. It has no color, taste, or smell and gives you no warning of its presence, but it is NOT harmless

If your answer indicated that carbon monoxide distorts your judgment, give yourself                                     **2**
(Victims of carbon monoxide poisoning may make irrational attempts at escape, or may waste vital minutes saving items of little or no value. People who have been in a burning building for some minutes should be watched, to be sure they do not go back into the fire.)

If your answer indicated that carbon monoxide disturbs your coordination (making simple escape efforts, such as unlocking a window difficult, or impossible), give yourself                                     **2**

---

**Question 10.** Assume you plan to hang by your hands from a window ledge and then drop to the earth below. Estimate in feet the distance you could drop and still have a 50:50 chance of surviving without serious injury?

Score yourself in accordance with the following table:

If your answer was —
Less than 20 feet: score                                     **3**
More than 20 feet, but less than 25 feet: score                                     **1**
More than 25 feet, but less than 35 feet: score                                     **0**
More than 35 feet, but less than 50 feet: subtract 2 points.
More than 50 fee: subtract 3 points.

Add 1 point if you have had training as a parachute jumper.

Subtract 1 point if you are over 50 years of age, unless your answer was under 15 feet.

---

**Question 11. (a)** What is the reason for having fuses in an electrical circuit?

If your answer indicates that the purpose of a fuse is to prevent a fire (by "blowing" before the wires can overheat when too much of a load is put on them), give yourself     **3**

---

**Question 11. (b)** What strength fuse should be used in an ordinary lighting circuit?

If your answer advised a 15 amp. fuse, give yourself     **3**

If your answer advised a 30 amp. fuse, subtract 3 points.

---

**Question 12.** What number should you dial to report a fire by telephone, and how should you report it?

If your telephone area is on the 911 emergency system, and you wrote down 911, or
If you gave the correct number for your local fire department, give yourself     1

If you said you would give the location of a fire slowly and clearly, give yourself     1

If you said that you would stay on the line to give additional information requested by the fire department, if you could do so safely, give yourself     1

If the number you called (police or "operator") would result in a delay in transmitting the message to the fire department, give yourself only     1

If you gave the wrong number, either for the fire department, or the police, or left the question unanswered, subtract 3 points.

**Question 13.** When is an electric cord dangerous? (give at least two examples)

If you listed any two of the following, give yourself    **3**
  When it is frayed;
  When the insulation has worn off;
  When it is wet;
  When it is under a rug (where repeated walking on it
    may break the insulation);
  When it is run over a nail (where the insulation may
    break at the bend);
  When it is run through a doorway (where closing the
    door may cause a break in the insulation);
  When it is pulled out of a wall socket by the wire,
    instead of by holding onto the plug, so there is danger
    of one of the wires coming loose and touching the
    other; and
  When nails are driven into it.

---

**Question 14.** When is a double plug dangerous?

If your answer included: When it is broken or when it is
  wet, give yourself    **1**

If it included when it is overloaded, (by having many
  appliances plugged into it), give yourself    **3**

---

**Question 15.** What should you do if you discover a large fire in your basement?

If your answer included:

  Shutting the basement door, give yourself    **3**

  Calling the fire department, give yourself    **3**

  Getting everyone, including yourself, out of the house,
    give yourself    **3**

If your answer included trying to fight a basement fire
  yourself, subtract 2 points. If it included fighting the fire
  yourself, without having notified the fire department,
  subtract 3 points, instead of 2.

**Question 16.** If you are trying to light a gas oven or burner and the first match goes out too soon, what should you do?

If your answer included turning off the gas before lighting
    a second match (so that explosive quantities of gas
    would not accumulate in the oven or burner to be set off
    by the second match), give yourself     **3**

If you made sure the first match was completely out, by
    breaking it or touching the tip, before discarding it, give
    yourself     **1**

---

**Question 17.** What is meant by "spontaneous combustion" or "spontaneous ignition"?

If your answer described the ignition of substances (such
    as wet newspapers, oily rags, paint-covered wipe cloths,
    and damp hay), which generate their own heat and
    ignite without the application of an external source, give
    yourself     **2**

---

**Question 18.** How should you store oily or greasy rags?

If you answered that they should not be kept
    or
If you said they should be kept in a closed metal
    container, give yourself     **3**

---

**Question 19.** Why should gasoline be stored only in metal cans with self-closing caps?

If you answered:
    To prevent fires, give yourself     **3**
    Because metal cans will not break readily, give yourself     **3**

If you answered to prevent fumes from spreading across
    the floor (which may be ignited by a spark, cigarette, or
    hot furnace), give yourself     **3**

---

**Question 20.** Should you put out an electric fire with water?

If you answered no, give yourself     **3**

---

Add up your points to determine your fire safety score.
Maximum possible score = 100 (101 for parachute jumper).

# Appendix II

## Glossary

**Alarm signal:**
A loud noise that warns or alerts

**Arson:**
The malicious or fraudulent burning of property.

**Circuit Breaker:**
A switch that automatically interrupts an electric circuit under an infrequent abnormal condition.

**Combustible:**
Capable of burning.

**Conflagration:**
A large disastrous fire that crosses natural fire barriers such as streets, and usually involving buildings in more than one block. It may also include forest fires and fires involving whole communities. It does not refer to group fires confined to a single group of buildings.

**Fireproof:**
To proof against or resistant to fire.

**Fire-resistant:**
A relative term, indicating that a material or structure withstands fire; usually with a numerical rating or modifying adjective, eg., "fire resistant up to 2 hours as measured on the Standard Time-Temperature Curve."

**Fire-retardant:**
A term generally used to refer to materials that are partially combustible, and have been treated or covered to prevent or retard ignition or the spread of fire

under the conditions for which they are designed.

**Firestopping:**
Using material to close open parts of a building for preventing the spread of fire.

**Fire Zone:**
An area stripped of vegetation to prevent the possibility of brush and forest fires.

**Flammable:**
Capable of being easily ignited and of burning with extreme rapidity. Identical in meaning with inflammable, and used interchangably.

**Flashover:**
The phenomenon of a slowly developing fire (or radiant heat source) producing radiant energy at wall and ceiling surfaces. The radiant feedback from those surfaces gradually heats the contents of the fire area, and when all the combustibles in the space have become heated to their ignition temperature, simultaneous ignition occurs as from a pilot ignition source.

**Photographic Inventory:**
A documentation of personal property by taking photographs of the belongings to be used in case of fire or theft for insurance claims.

**Smoke Detector:**
A device that sets off an alarm when visible or invisible particles of combustion come in contact with it.

**Space Heaters:**
An electric device that imparts heat to the surrounding area.

**Sprinkler System:**
A system for protecting a building against fire by means of overhead pipes which convey an extinguishing fluid (as water) to heat-activated outlets.

**Telephone Dialer System:**
A device, which when actuated by an alarm system, seizes a telephone line, dials a number or numbers, and transmits an appropriate emergency message or messages to each party called.

**Underwriter's Laboratories:**
An independent, non-profit organization conducting tests of products for public safety and/or security. A U.L. label is affixed to those products that have passed U.L. tests.

# Index